Quantitative Literacy Through Games and Gambling

This book was developed to address a need. Quantitative Literacy courses have been established in the mathematics curriculum for decades now. The students in these courses typically dislike and fear mathematics, and the result is often a class populated by many students who are unmotivated and uninterested in the material.

This book is a text for such a course; however, it is focused on a single idea that most students seem to already have some intrinsic interest in and is written at an accessible level. It covers the basic ideas of discrete probability and shows how these ideas can be applied to familiar games (roulette, poker, blackjack, etc.). The gambling material is interweaved through the book and introduced as soon as the necessary mathematics has been developed. Throughout, mathematical formalism and symbolism have been avoided, and numerous examples are provided.

The book starts with a simple definition of probability, goes through some basic concepts like combining events and expected value, and then discusses some elementary mathematical aspects of various games. Roulette is introduced very early on, as is the game of craps, which requires some knowledge of conditional probability. Other games like poker, blackjack, and lotteries, whose study requires some rudimentary combinatorics, come shortly thereafter. The book ends with a brief introduction to zero-sum games, with some attention paid to the use of these ideas in studying bluffing.

In addition to discussion of these traditional games, the author motivates probability by talking about a few applications in legal proceedings that illustrate how mathematics has been misused in the courtroom. There is also a discussion of the Monty Hall problem, a nonintuitive result in probability that has an interesting and colorful history.

Hopefully, students studying from this text will find that mathematics is not as horrible as they have always thought and offers some interesting applications in the real world. This should perhaps be the goal of any quantitative literacy course.

Textbooks in Mathematics
Series editors: Al Boggess, Kenneth H. Rosen

An Introduction to Optimization with Applications in Data Analytics and Machine Learning
Jeffrey Paul Wheeler

Encounters with Chaos and Fractals, Third Edition
Denny Gulick and Jeff Ford

Differential Calculus in Several Variables
A Learning-by-Doing Approach
Marius Ghergu

Taking the "Oof!" Out of Proofs
Alexandr Draganov

Vector Calculus
Steven G. Krantz and Harold Parks

Intuitive Axiomatic Set Theory
José Luis García

Fundamentals of Abstract Algebra
Mark J. DeBonis

A Bridge to Higher Mathematics
James R. Kirkwood and Raina S. Robeva

Advanced Linear Algebra, Second Edition
Nicholas Loehr

Mathematical Biology: Discrete and Differential Equations
Christina Alvey and Daniel Alvey

Numerical Methods and Analysis with Mathematical Modelling
William P. Fox and Richard D. West

Business Process Analytics
Modeling, Simulation, and Design
Manuel Laguna and Johan Marklund

Mathematics and Gambling
A Course in Quantitative Literacy
Mark Hunacek

https://www.routledge.com/Textbooks-in-Mathematics/book-series/CANDHTEXBOOMTH

Quantitative Literacy Through Games and Gambling

Mark Hunacek

CRC Press
Taylor & Francis Group
Boca Raton London New York

CRC Press is an imprint of the
Taylor & Francis Group, an **informa** business

A CHAPMAN & HALL BOOK

First edition published 2025
by CRC Press
2385 NW Executive Center Drive, Suite 320, Boca Raton FL 33431

and by CRC Press
4 Park Square, Milton Park, Abingdon, Oxon, OX14 4RN

CRC Press is an imprint of Taylor & Francis Group, LLC

© 2025 Mark Hunacek

Reasonable efforts have been made to publish reliable data and information, but the author and publisher cannot assume responsibility for the validity of all materials or the consequences of their use. The authors and publishers have attempted to trace the copyright holders of all material reproduced in this publication and apologize to copyright holders if permission to publish in this form has not been obtained. If any copyright material has not been acknowledged please write and let us know so we may rectify in any future reprint.

Except as permitted under U.S. Copyright Law, no part of this book may be reprinted, reproduced, transmitted, or utilized in any form by any electronic, mechanical, or other means, now known or hereafter invented, including photocopying, microfilming, and recording, or in any information storage or retrieval system, without written permission from the publishers.

For permission to photocopy or use material electronically from this work, access www.copyright.com or contact the Copyright Clearance Center, Inc. (CCC), 222 Rosewood Drive, Danvers, MA 01923, 978-750-8400. For works that are not available on CCC please contact mpkbookspermissions@tandf.co.uk

Trademark notice: Product or corporate names may be trademarks or registered trademarks and are used only for identification and explanation without intent to infringe.

Library of Congress Cataloging-in-Publication Data
Names: Hunacek, Mark, author.
Title: Quantitative literacy through games and gambling / Mark Hunacek.
Description: First edition. | Boca Raton, FL : CRC Press, 2025. | Series: Textbooks in mathematics | Includes bibliographical references and index.
Identifiers: LCCN 2024017769 | ISBN 9781032659152 (hbk) | ISBN 9781032633923 (pbk) | ISBN 9781032659145 (ebk)
Subjects: LCSH: Probabilities. | Chance. | Gambling--Mathematics. | Games of chance (Mathematics)
Classification: LCC QA273 .H86 2025 | DDC 519.2--dc23/eng/20240603
LC record available at https://lccn.loc.gov/2024017769

ISBN: 9781032659152 (hbk)
ISBN: 9781032633923 (pbk)
ISBN: 9781032659145 (ebk)

DOI: 10.1201/9781032659145

Typeset in Palatino
by KnowledgeWorks Global Ltd.

This book is dedicated to Leslie, Adrienne, and Sofia—the three most important women in my life

Contents

About the Author ix
Preface xi

1. **Sample Spaces and Events** 1
   1.1 Basic Set Theory 1
   1.2 Sample Spaces 3
   1.3 Events 5
   1.4 Probability 7
   1.5 The Complement of an Event 10
   1.6 Expected Value 12
   1.7 Odds 16

2. **Roulette** 18
   2.1 Rules of the Game 18
   2.2 Some Roulette Calculations 20
   2.3 Roulette "Systems" 23

3. **Conditional Probability and Independence** 26
   3.1 Definition of Conditional Probability 26
   3.2 Law of Total Probability 30
   3.3 Independent Events 33
   3.4 The Monty Hall Problem 36

4. **Craps** 39
   4.1 Rules of the Game 39
   4.2 Analysis of the Shooter's Game 41
   4.3 Other Bets 44

5. **Counting Large Sets: An Introduction to Combinatorics** 47
   5.1 Two Counting Rules 47
   5.2 Permutations and Combinations 53

6. **Poker** 58
   6.1 Poker Hands and Their Probabilities 58
   6.2 Video Poker 62
   6.3 Texas Hold 'Em 66

7. **Lotteries and Keno** 68
   7.1   Lotteries and Powerball 68
   7.2   Keno 72

8. **Blackjack** 75
   8.1   Rules of the Game 75
   8.2   Basic Blackjack Calculations 77
   8.3   Card Counting 80

9. **Farkle** 82
   9.1   Rules of the Game 82
   9.2   Some Probability Calculations 84
   9.3   Should You Risk Another Roll? The Probability
         of Farkling 86

10. **An Introduction to Game Theory** 89
   10.1   Introduction and Basic Definitions 89
   10.2   Zero-sum Games: Domination 92
   10.3   Zero-sum Games: Saddle Points 95
   10.4   Zero-sum Games: No Saddle Points 96
   10.5   Solving 2×2 Zero-sum Games 99
   10.6   A Simplified Poker Game 102

Index 105

About the Author

Mark Hunacek received his Ph.D. in mathematics from Rutgers University and met his wife who also received her Ph.D. in mathematics the same year. Faced with the familiar "two-body problem", which was more of an issue in 1978 than it is now, he went to law school and then spent almost three decades practicing law. After retiring from the practice of law, things came full circle, and he was lucky enough to be offered a position in the mathematics department at Iowa State University, where, as one of his responsibilities, he redesigned and oversaw the two quantitative literacy courses offered there, one of which inspired this textbook. In 2021, he retired and became a Teaching Professor Emeritus.

Preface

When I joined the mathematics department at Iowa State University, I was asked by the then-chair to revamp and become course coordinator of our two "quantitative literacy" courses. These courses were designed for students who needed to fulfill a mathematics requirement but who did not have the interest or ability to take calculus. One of these courses, Math 104, had been previously taught as a finite mathematics course, covering a hodgepodge of material (probability, combinatorics, matrices, etc.) that the students found unmotivated and uninteresting. The result was a course that was unpopular with both the students and the faculty.

It seemed to me that one way to help fix the course was to focus on a particular topic, preferably one that the students already had some intrinsic interest in and that was "real-world relevant". I chose games and gambling as this topic, both because it seemed like the kind of thing students would enjoy, and also because there were several books out that could serve as a text, including Gould's *Mathematics in Games, Sports and Gambling* and Packel's *The Mathematics of Games and Gambling*. As I continued to supervise these courses, other books on the subject were published: for example, Bollman's *Basic Gambling Mathematics: The Numbers Behind the Neon* and Taylor's *The Mathematics of Games: An Introduction to Probability*.

Although these are all excellent books, the students in my classes said they found them too "math-y": more suitable for, say, an honor's seminar than a freshman quantitative literacy course for students who generally hated or (and in many cases) feared mathematics. Ultimately, I wound up using a "custom book" consisting of a lengthy chapter on probability from a finite math book along with notes written by me. It was from these notes that the idea to write this book developed.

The purpose of this book is to develop basic discrete probability at as low a level as possible while still actually teaching some mathematics, and to interweave, throughout this discussion, applications of these mathematical ideas to various games of chance, most of them played in casinos. While this book could possibly be used in a seminar, its primary intended audience is the kind of student who finds himself or herself dragged, kicking and screaming, into a required course, often as a freshman.

As the table of contents makes clear, we start with a simplified definition of probability (under the assumption of a finite sample space with all elements in it equally likely), go through some basic concepts like combining events and expected value, and introduce mathematical discussions of various games as soon as enough mathematics has been developed to do so. Thus, roulette, which requires little mathematics other than the definition of probability and the concept of expected value, is introduced very early on, as is the game of craps, whose analysis requires some knowledge of conditional probability. Other games like poker, blackjack, and lotteries, whose study requires some rudimentary combinatorics (because of the size of the relevant sample spaces), come shortly thereafter. The book does not delve too deeply into these games but does give an idea of how much per dollar the casino can be expected to obtain for each one, and how probability can be used to determine various probabilities that might arise during gameplay. I have also included a chapter on Farkle, a game not played in a casino, simply because I find the game rather addictive and many of my students have also played it on Facebook. The book ends with a brief introduction to zero-sum games, with some attention paid to the use of these ideas in studying bluffing.

Throughout, mathematical terminology and notation have been kept to a minimum. Words like "theorem" and "proof" are assiduously avoided (as are, of course, actual proofs), and so is mathematical notation (like the sigma notation for sums). The axioms of probability are never mentioned; the probability of an event A in a sample space S is simply defined (given our simplifying assumptions) as the number of elements in A divided by the number of elements in S. Conditional probability is not defined by means of a formula but rather in terms of "adjusted sample spaces". I found that the students found this approach to be much more intuitively understandable.

In addition to discussion of these traditional games, I have also tried to motivate probability by talking about a few applications in legal proceedings; the Sally Clark case in England, for example, and the famous Collins case in California, both of which provide some illustration of how mathematics has been misused in the courtroom, are discussed. There is also a discussion of the Monty Hall problem, a nonintuitive result in probability that has an interesting and colorful history. Other interesting and counterintuitive results in basic discrete probability, like the famous birthday problem, are also discussed.

It is my hope that a student studying from this text will find that mathematics is not as difficult and horrible as he or she has always believed it to be and can actually find some interesting applications in the real world. That, I think, should be the ultimate goal of any quantitative literacy course.

1
Sample Spaces and Events

In this chapter, we define probability, at least in the simple context in which we will be using it, and give some examples. We will then use the concept of probability to discuss "expected value", an idea that is crucial to any understanding of the mathematics of gambling. The term "probability" is defined in terms of sets, so we begin with a very brief overview of some basic terminology and ideas of set theory.

1.1 Basic Set Theory

Set theory is one of those branches of mathematics that can be discussed at varying levels of sophistication. Fortunately for us, we do not need to get involved in any mathematical subtleties and can study it at a very elementary level.

For our purposes, a *set* is simply a collection of objects; these objects are called *elements* of the set. One way to define a specific set is by a verbal description; for example, "the set of the first five letters of the English alphabet". In some cases, we can also define a set by simply listing the elements in it. For the example just given, we could also write the set as {A, B, C, D, E}. (It is common to put the elements of the set inside curly brackets.) Of course, if the number of elements in a given set is very large, or infinite, then it is inconvenient at best, and impossible at worst, to actually list the elements in it. We cannot, for example, specifically list all the elements in the set of positive integers, though we can write {1, 2, 3, ...}, with the dots at the end signifying that the list goes on forever. On the other hand, if we are dealing, say, with the "set of all real numbers", then it is literally impossible to list the elements. In this book, all the sets we encounter will be finite, but some will be fairly large. (Would the reader like to guess, for example, roughly how many elements are in the set of all possible 5-card poker hands? Later on in the book, we will learn techniques for computing this number quickly and easily. The answer, by the way, turns out to be 2,598,960.)

We will generally denote sets by uppercase letters of the English alphabet; for example, S = {A, B, C, D, E}. Two sets X and Y are *equal* (denoted as X = Y) if they have precisely the same elements. We use the mathematical symbol ∈ to mean "is an element of"; so, for this set S, it would be correct to write C ∈ S.

We use the symbol $\notin$ to mean "is not an element of"; so, in this case, it would be correct to write $F \notin S$ (or, for that matter, $5 \notin S$).

If A and B are sets, and every element of A is also an element of B, then we say that A is a *subset* of B, and we write $A \subseteq B$. It should be clear from the definition that $A \subseteq A$: after all, every element of A is an element of A. It should also be clear that two sets A and B are equal when and only when $A \subseteq B$ and $B \subseteq A$. If $A \subseteq B$ but A and B are not equal, then we say that A is a *proper subset* of B, and write $A \subset B$. For example, $\{1, 2, 3\} \subseteq \{1, 2, 3, 5\}$, and also $\{1, 2, 3\} \subset \{1, 2, 3, 5\}$. The symbol $\nsubseteq$ means "is not a subset of"; to say $A \nsubseteq B$, therefore, simply means that there is at least one element of A that is not an element of B. So, if $A = \{2, 4, 6, 10, 14\}$ and $B = \{1, 2, 3, 4, 5, 6\}$, then we have $A \nsubseteq B$ and $B \nsubseteq A$. (To the reader: stop at this point and make sure you can explain this to yourself.)

Consider the set described verbally as "the set of all negative numbers greater than 1". Obviously, there are no numbers satisfying this condition; so, this set has no elements in it at all. The set with no elements is called the *null set* and is denoted by $\emptyset$. It may not be obvious, but the null set is a subset of *every* set. Here's why: suppose that, for some set A, if the case were $\emptyset \nsubseteq A$, then by the comments in the preceding paragraph, this would mean there was some element in $\emptyset$ that was not an element of A. But that's impossible because there are no elements in $\emptyset$ at all. Since $\emptyset \nsubseteq A$ is not possible, it must be the case that $\emptyset$ is a subset of A.

Two sets A and B can be combined. The *union* of A and B, denoted as $A \cup B$, is simply the set of all elements that are either in A *or* B (or both). The *intersection* of A and B, denoted as $A \cap B$, is the set of all elements that are in both A *and* B. For example, if $A = \{2, 4, 6, 10, 14\}$ and $B = \{1, 2, 3, 4, 5, 6\}$, then $A \cup B = \{1, 2, 3, 4, 5, 6, 10, 14\}$ and $A \cap B = \{2, 4, 6\}$. (Note that when taking this union, we did not write down elements like 2 and 4 twice, even though they appear in both A and B: an element is either in a set or not in it; it does not make sense to ask "how many times is it in the set?".) Finally, it is, of course, perfectly possible for A and B to have no elements in common, in which case we have $A \cap B = \emptyset$.

Exercises

1.1. If $A = \{2, 4, 6, 8, 10, 12, 14\}$ and $B = \{5, 10, 15, 20\}$, determine $A \cup B$ and $A \cap B$.
1.2. Is $\{\emptyset\}$ the same as $\emptyset$? Explain.
1.3. Write down all subsets of $\emptyset$, $\{1, 2\}$ and $\{1, 2, 3\}$.
1.4. On the basis of your answers to the previous question, guess (but do not attempt to justify) a formula for the number of subsets of a set with n elements, where n is any positive integer.

1.5. (This question assumes that the reader is familiar with an ordinary deck of playing cards. Those who are not can read the second paragraph of the next section, or look up the Wikipedia entry "Standard 52-card deck".) Let A be the set of red cards (hearts and diamonds) in a normal 52-card deck. Let B be the set of queens in this deck. How many elements are in $A \cup B$ and $A \cap B$?

1.6. If a coin is flipped, it will show either a head (H) or a tail (T). (We ignore the very unlikely possibility of it landing on its side.) Write down, in a convenient notation, the set of all outcomes if the coin is flipped three times.

1.7. If $A \cup B = \emptyset$, what can you say about the sets A and B? Explain.

1.2 Sample Spaces

Whenever an experiment (rolling dice, picking a card, flipping a coin, etc.) is conducted, there will be a number of ways in which the experiment can turn out. The set consisting of these possible outcomes is called the *sample space* of the experiment. So, for example, if we flip a coin, a natural sample space would be {H, T}, where, of course, H and T stand for "heads" and "tails", respectively. If we roll a single die, the sample space might be {1, 2, 3, 4, 5, 6}.

Some of the experiments we will discuss in this book involve a deck of cards. Although we assume that most readers have grown up playing card games, we will give a brief description of a deck of cards for those who might not be familiar with them. An ordinary deck contains 52 cards, each one having a *rank* and a *suit*. The *rank* of a card is a number from 2 to 10, or one of Jack, Queen, King, and Ace. There are four cards for each of these 13 ranks—a heart, diamond, spade, or club, denoted by the symbols ♥, ♦, ♠, and ♣, respectively. The heart and diamond cards are red; the others are black. With this as background, we see that if a card is drawn from an ordinary deck, the sample space will have 52 elements in it. If the red cards are removed from a deck and then a card is drawn from them, the sample space will have 26 elements in it.

Note that in these examples, the sample spaces are both finite (i.e., there are a specific positive number of elements in each one), and also, in each one, the outcomes are equally likely to occur. In this book, *we will only consider situations where these two conditions are satisfied.* (This standing assumption provides a huge simplification when it comes to defining the probability of an event.) So, for example, if our experiment consisted of asking a friend to "pick a number—any number", then that particular experiment would *not* have a finite number of outcomes and would thus be out of the purview of this text. Defining probability becomes a lot more difficult in a situation like this; fortunately, we won't have to worry about these difficulties.

Let us look at some other examples. If we flip a coin twice, a reasonable sample space might be {HH, HT, TH, TT}, where, in what we hope is an obvious notation, HT means "heads on the first toss, tails on the second" (with similar meanings attached to the remaining symbols). If we flip a coin three times, the set {HHH, HHT, HTH, HTT, THH, THT, TTH, TTT} makes a reasonable sample space.

Moving from coins to dice, suppose we roll two dice. Again, to avoid technical complications, we assume the dice are *distinguishable*—let us say one is green and the other is red. If we are interested in the number of dots shown on each of these dice, we could use 12 to mean "1 on the green, 2 on the red", but this risks confusion with the number 12. Instead, we will use the notation (1, 2) to reflect this occurrence. With that in mind, the elements of a reasonable sample space might be

(1, 1) (1, 2) (1, 3) (1, 4) (1, 5) (1, 6)
(2, 1) (2, 2) (2, 3) (2, 4) (2, 5) (2, 6)
(3, 1) (3, 2) (3, 3) (3, 4) (3, 5) (3, 6)
(4, 1) (4, 2) (4, 3) (4, 4) (4, 5) (4, 6)
(5, 1) (5, 2) (5, 3) (5, 4) (5, 5) (5, 6)
(6, 1) (6, 2) (6, 3) (6, 4) (6, 5) (6, 6)

It will soon become apparent that our primary interest in sample spaces (for computing probabilities, anyway) is in the *size* of the sample space. The examples above help illustrate an important tool in determining the number of elements in the sample space, even when it is impracticable to write them all down explicitly. Note that when we flipped a coin (with two outcomes) once, the sample space had 2^1 elements; and when we flipped it twice, the sample space had 2^2 elements; when we flipped it three times, there were 2^3 elements in the sample space. And when we rolled two dice (with six outcomes per die), the number of elements was 36, or 6^2. These outcomes are all special cases of an important counting principle that we will discuss in more detail later in the book but which we can at least state and start to use now. It is called the *multiplication principle* for counting:

> if a first task can be done in M possible ways and, regardless of how the first task was done, a second task can be done in N possible ways, then there are M × N ways to do the two tasks, one after the other.

This principle extends from two tasks to three or more in a natural way. Here is a simple example: a menu offers three appetizers, four entrées, and four desserts. The number of ways to order an appetizer, entrée, and dessert is the product of 3, 4, and 4, or 48.

Another example: suppose we flip a coin ten times. The first task is to record the result on the first flip (two possible choices). Ditto for the second

Sample Spaces and Events

through tenth flips. So, the total number of ways to record the results in all ten flips is $2 \times 2 \times 2 \ldots \times 2$ (ten times), or 2^{10} (= 1,024). So, we know how many elements are in this sample space, even though we cannot reasonably write them all out.

Exercises

1.8. If a fair coin is flipped and then a die is rolled, what is a reasonable sample space?
1.9. Suppose two dice are rolled, but we are interested only in the sum of the dice. Is {2, 3, 4, 5, 6, 7, 8, 9, 10, 11, 12} a valid sample space for our purposes? Why or why not?
1.10. Three balls (one red, one white, and one black) are in a box. John picks one out at random, notes its color, and discards it. He then picks another ball from the two remaining. Write down a reasonable sample space for this experiment. (Use the letters R, W, and B to denote the balls.)
1.11. Now re-do problem 1.10, this time assuming John replaces the ball in the box before selecting a new one.
1.12. The Aces from a deck of cards are put in front of you, face down. You select one at random. Using any notation that you find convenient, write down a reasonable sample space for this experiment.

1.3 Events

The elements of the sample space are sometimes called *elementary events*. Each elementary event, therefore, is an outcome of some experiment—some way in which the experiment can turn out. In real life, though, people are often interested in events that can occur in a number of different ways. In the dice game of craps, for example, a pair of dice is rolled, but the players are interested not so much in the individual numbers that appear but in their sum, and any particular sum can sometimes appear in more than one way. To get a sum of three, for example, we can have the first die show a 1 and the second a 2, or vice versa; there are thus two possible ways this can occur. A sum of 7 can occur in six ways: (1, 6), (2, 5), (3, 4), (4, 3), (5, 2), and (6, 1). We can thus think of the event A ("the sum of the dice is 7") as the set {(1, 6), (2, 5), (3, 4), (4, 3), (5, 2), (6, 1)}.

As another example, suppose we flip a (fair) coin three times. Let A be the event "there are exactly two heads". Then the sample space for this experiment has eight elements, and A can be written as the set {HHT, HTH, THH}.

These examples lead to the definition of an *event* in a sample space as a subset of that sample space. (The event may be described in words, but ultimately we think of it as a collection of elements of the sample space.) If an experiment results in an outcome that is an element of A, then we say "event A occurs". Events are the things to which we will eventually assign probabilities.

It will be recalled from Section 1.1 that the null set $\emptyset$ is a subset of every set. Therefore, for any sample space S, $\emptyset$ is an event in S. Since the null set has no elements in it, it corresponds to an event that does not occur—i.e., an *impossible event*. (For example, in the sample space of 36 elements that corresponds to rolling two dice, consider the event "the sum of the dice is 13".) At the other extreme, the sample space S is a subspace of itself; this corresponds to an event that must occur. (For example, "the sum of two dice is at least 2".) We call this the *certain event*. Our intuition tells us that the empty set should have small probability, while the certain event should have maximum possible probability. We will confirm our intuition in the next section, where we define "probability".

If A and B are events on a sample space S, then it makes sense to talk of two other events: "A and B both happen" and "A or B happens". The set that corresponds to the first of these events is nothing more than $A \cap B$, the intersection of A and B; we will often denote it by AB in this book to avoid unnecessary use of symbols. Likewise, the set corresponding to the second of these events is $A \cup B$. An example should make this clear. Suppose we roll a single die. If A is the event "the die shows an even number" and B is the event "the die shows a number greater than 3", then in terms of sets, A = {2, 4, 6} and B = {4, 5, 6}. Now consider the event "A and B both happen". The only way A and B can both occur is for the die to show an even number greater than 3—i.e., the die shows a 4 or 6. So, "A and B" corresponds to the set {4, 6}, which of course is $A \cap B$. Likewise, "A or B" will occur if and only if the die is either even (i.e., in A) or greater than 3 (i.e., in B), and this is precisely the definition of $A \cup B$.

Exercises

1.13. Consider the experiment described in exercise 1.10. Let A be the event "one of the balls selected is white". Write out this event as a set.

1.14. Consider the experiment described in exercise 1.11. Let A be the event "one of the balls selected is white". Write out this event as a set.

1.15. In the previous exercise, let B be the event "both balls have the same color". Write out this event as a set.

1.16. We have already noted that if we roll two distinguishable dice, the sample space has 36 elements. If A is the event "the sum of the dice is greater than 10", write out A as a set.

1.17. (Continuation of previous problem.) Let B be the event "there is at least one 6". Write out B as a set. Then write out the event "both A and B occur" as a set.
1.18. Suppose a fair coin is flipped four times. Write out the event "there are at least three consecutive heads" as a subset of the sample space.

1.4 Probability

We have now reached the point where we can define the probability of an event A in a sample space S. Recall that, intuitively, the probability of an event is a reflection of its likelihood of occurring, so that the larger the set A is, the higher its probability should be. It would therefore make sense to define the probability of A in terms of the number of elements in A. To simplify the notation in what follows, we will denote by #A the number of elements in a finite set A.

> **Definition 1.4.1.** If A is an event in a sample space S (where S is finite and all the elements of S are equally likely), then the *probability* of A, denoted P(A), is $\#A/\#S$.

Note that this definition makes sense because both A and S are finite sets. (This is one of our assumptions about sample spaces.) It also captures our intuitive notion of probability, since the larger A is, the larger $\#A/\#S$ is. It is useful to think of this definition as "favorable divided by total" in that the fraction represents the number of ways A can occur divided by the total number of possible outcomes.

Let us give some examples of the probability concept. Most people, if asked what the probability that a flipped (fair) coin will come up heads, will immediately say ½. This is also the answer we get (as it had better be!) using the definition: the sample space {H, T} has two elements in it, only one of which, H, corresponds to the given event.

Here is a less trivial example: if we flip a coin three times, what is the probability that there are exactly two heads? Recall that the sample space here is {HHH, HHT, HTH, HTT, THH, THT, TTH, TTT}, an eight-element set. Of the eight elements in this set, only three correspond to a "favorable" outcome: HHT, HTH, and THH. The answer is thus ⅜. Question for the reader: what is the probability that all three flips show the same face of the coin?

Let's do a dice example (one that will be useful for us when we study craps later in the book): if two (distinguishable, as always) dice are rolled, what is the probability that their sum is 7? If we refer to the sample space whose

36 elements are written out in Section 1.2, we see that of the 36 possible outcomes, six of them are "favorable": (1, 6), (2, 5), (3, 4), (4, 3), (5, 2), and (6, 1). Dividing 6 by 36 yields the answer: ⅙.

Two events in any sample space S that deserve special mention are the null set $\emptyset$ and the entire sample space S. Using definition 1.4.1, it is easy to see that the probabilities of these events are 0 and 1, respectively. In general, if A is any event, then P(A) is a number that lies between 0 and 1. So, it never makes sense to say that an event has probability 2 or -4.

Probability can be a very counter-intuitive area of mathematics; it abounds with results that many people have trouble believing. For example, suppose a fair coin has just been flipped four times and has come up heads each time. Many people would assume that the probability that the fifth flip will also come up heads is less than ½, just because it is very unlikely that a fair coin will show the same face five times in a row. In fact, however, that is false: the probability is exactly ½. (Of course, when you think about it, this result isn't all that hard to believe after all; the coin has no memory.) We will see other counter-intuitive results later in the book.

We next address the question of how probability behaves with respect to combining events. If A and B are events, a natural question to ask is how $P(A \cup B)$ relates to $P(A)$ and $P(B)$. Your instinct may tell you that $P(A \cup B) = P(A) + P(B)$, but a moment's thought should disabuse you of this notion. For example, if S = {1, 2, 3, 4, 5, 6}, A = {1, 2}, and B = {1, 3}, then it is easy to verify (do so!) that $P(A \cup B) \neq P(A) + P(B)$. The reason that this is the case is because $\#(A \cup B) \neq \#A + \#B$, and the reason that *this* is so is because when we lump together the elements of A and those of B, one of these elements, namely 1, gets counted twice. So, $\#(A \cup B)$ misses being equal to $\#A + \#B$ precisely because every element in both A and B gets counted twice in $\#A + \#B$ but only once in $\#(A \cup B)$. So, to obtain a correct equation, we must subtract the number of overcounts: the correct formula, therefore, is $\#(A \cup B) = \#A + \#B - \#(A \cap B)$. Dividing by #S gives the formula for probabilities:

$$P(A \cup B) = P(A) + P(B) - P(A \cap B).$$

Of course, when A and B have no elements in common, then $P(A \cap B) = 0$ and the "intuitive" formula holds.

Next, let us consider $P(A \cap B)$. Here again, instinct may suggest that this is equal to P(A)P(B), but again, our instinct is wrong. Indeed, the same example given above also shows this, a point we encourage the reader to verify for himself or herself. The question of when $(A \cap B)$ *does* equal P(A)P(B) is an interesting one that we defer until later in this text. Suffice it to say now that a failure to properly understand this fact has led to some horrible mistakes in courtrooms, including wrongful convictions for homicide. We will elaborate on this topic in Section 3.5.

Sample Spaces and Events

In closing this section, let us mention one other aspect of the concept of probability. Many people instinctively think of the probability of an event as expressing something about the long-term behavior of that event. For example, we know that the probability of coming up heads on a single flip of a fair coin is ½. For many people, this expresses the intuitive idea that if we flip a coin lots and lots of times, we would expect the total number of heads to be close to one-half the total number of flips. Such an interpretation, however, is not readily apparent from our definition of "probability", which, after all, says nothing about repeating events. Nevertheless, it turns out to be true and is reflected in a mathematical result called The Law of Large Numbers. Not only is the proof of this result far beyond the scope of this book, but the very *statement* of the law is too. We can, however, offer a somewhat imprecise but nonetheless suggestive formulation of this result: *If an experiment is conducted N times, where N is a very large number, and if A_N denotes the number of times A occurs in these N experiments, then the ratio A_N/N is very close to P(A), the probability of A.*

In advanced courses in mathematics, the notion of being "very close to" is expressed by the concept of "limit", a topic that is taught in calculus classes. The proof of this result is therefore typically reserved for advanced courses where calculus is a prerequisite. Although we won't give it here, we should at least recognize that this result is very important for casinos. In the famous casino game of roulette, for example, we will see that the probability of a gambler losing a bet on the color "red" is roughly 0.526, so the probability that the casino wins is slightly more than ½. The Law of Large Numbers tells a casino that although it may lose an individual red bet, in the long run, it will win more than half of them. Specifically, if a million bets are made, the casino can reasonably expect to win around 526,000 of them.

Exercises

1.19. Determine the probability of the event described in exercise 1.12.
1.20. Determine the probability of the event described in exercise 1.13.
1.21. Determine the probability of the event described in exercise 1.14.
1.22. Determine the probability of the event described in exercise 1.15.
1.23. A fair coin is flipped three times and comes up heads each time. Which do you think has greater probability, that the coin will come up heads on the fourth flip, or tails? Explain your answer.
1.24. You are dealt two cards from an ordinary, well-shuffled deck. What is the probability that the first card is an ace and the second is not? Explain.
1.25. A coin is flipped three times. Let A denote the event "all flips show the same side of the coin", B the event "the first flip is a head", C the event "the first and third flips match", and D the

event "heads and tails alternate". Compute the probability of each of the events A, B, C, D, AB, AC, AD, BC, BD, A or B, A or C, A or D, B or C, B or D, C or D.

1.26. John owns a special eight-sided die that contains the numbers 1 through 8 on it. It looks different than an ordinary six-sided die, of course, but is fair in the sense that, when rolled, any of the eight numbers has the same probability of coming up. If John rolls his special die, and Jane simultaneously rolls an ordinary die, what is the probability that John and Jane roll the same number? What is the probability that the sum of the two dice is 12? What is the probability that the sum of the two dice is 13?

1.27. You select, at random, a word from this sentence. What is the probability that you select a word with four letters in it?

1.5 The Complement of an Event

Consider the following question: if you flip a fair coin ten times, what is the probability that there will be at least one head and one tail? At first glance, this is a difficult problem to answer using the standard "favorable divided by total" definition of probability. Although we already know what the total number of possible outcomes is ($2^{10} = 1{,}024$), trying to compute the number of favorable outcomes seems like a formidable chore, simply because so many of the elements of the sample space contain both an H and a T. However, we can employ a trick.

Think of the sample space S as being written as the union of two non-overlapping sets: the set F of favorable outcomes and the set of unfavorable ones, which we will for the moment denote U. We want to compute P(F). Although this is hard to do directly, we do know that #S = #F + #U, since there is no overlap between F and U and, between them, they contain every element of S. Now, if we divide both sides of this equation by #S, we get 1 = #F/#S + #U/#S, or, putting it another way, 1 = P(F) + P(U). So, we don't really need to compute P(F) directly after all; it is enough to calculate P(U) instead and then subtract this number from 1.

And, in this case, calculating P(U) is a breeze. The only time an element of the sample space is not in F, if it corresponds to "all heads" (HHHHHHHHHH) or "all tails" (TTTTTTTTTT). In other words, there are two elements in U, so P(U) = $2/1{,}024$, and therefore P(F) = $1{,}022/1{,}024$.

This example, of course, is an illustration of a general principle. If A is any event in a sample space S, then the *complement* of A, denoted A^C, is the set of all elements in S that are *not* in A. For example, if S = {1, 2, 3, 4, 5, 6} and A = {1, 3}, then A^C = {2, 4, 5, 6}. Note also that the complement of S is ∅ (because there is no element of S that is not in S), and, likewise, the complement of ∅ is S (because every element of S is not in ∅).

Sample Spaces and Events

For any event A, A and A^C have, between them, all the elements of S, and none of these elements is in both of these sets. Therefore,

$$\#S = \#A + \#A^C.$$

Dividing by #S yields

$$1 = \#A/\#S + \#A^C/\#S$$

which in turn (referring to the definition of probability) gives

$$1 = P(A) + P(A^C).$$

This is, of course, the precise calculation we did a few paragraphs back, but then we were using the letters F and U instead of A and A^C.

This formula, relating the probability of an event to the probability of its complement, can be very useful because it is sometimes much easier to "count the complement" rather than try and figure out how many elements are in a given event. In general, this is likely to be the case for any event that uses phrases like "at least one" (as was the case with the example that started this section).

As a closing example, consider this: *if two distinguishable dice are rolled, what is the probability that there will be at least one 2 or 4 showing?* One way to solve this would be to look at the 36-element sample space (see Section 1.2) and manually count the ordered pairs that have at least one 2 or 4 in them. After a bit of tedium, you should arrive at the answer 20. Therefore, the answer to this question is $20/36$. A slicker way to do this, however, would be to consider the complimentary event, namely, the event that there are no 2s or 4s appearing. In this case, there are only four choices for the first die and four for the second, so the desired probability, by the multiplication rule of Section 1.2, is $16/36$. Since this is the probability of the complimentary event, the probability of the actual event must be $1 - 16/36 = 20/36$, as before.

Exercises

1.28. A student takes a ten-question multiple-choice exam. Each question has four possible answers, exactly one of which is correct. A student takes the test by guessing randomly on each question. What is the probability that this student gets at least one question correct?

1.29. Three distinguishable dice are rolled. What is the probability that at least one of them shows a 5?

1.6 Expected Value

The subject matter of this section, expected value, is of critical importance to the gambling industry. The owner of a casino knows that there will be times when a gambler wins big on the throw of some dice, the turn of a card, or the spin of a roulette wheel. The reason that the owner is not bothered by this example of gambler's luck is that he or she knows that in the long run, the casino will make money. The owner also knows exactly how many cents on each dollar bet the casino can expect to keep. This is the expected value of a particular bet.

To try and formulate the concept of expected value mathematically, let us first consider a simple example. Suppose that Alice and Bob enter into an agreement: they will flip a fair coin, and if the coin comes up heads, Alice will pay Bob a dollar; if the coin comes up tails, Bob will pay Alice a dollar. If Alice and Bob flip the coin a very large number of times (perhaps using computer simulation rather than actual flipping), most people would expect that they will each have close to zero dollars, simply because in the long run they will each win roughly half the time, and the wins will cancel out the losses.

We can think of this as a mathematical computation. Looking at things from Alice's perspective, on any one flip of the coin, he will win a dollar (+1) or lose a dollar (−1). Since wins and losses occur very close to half the time, Alice's ultimate outcome should be ½ (+1) + ½ (−1) = 0. The same calculation works from Bob's perspective as well.

As another example, consider the roll of a single die, and let us compute the expected number of spots that will appear. Since there are six possible outcomes (namely, of course, 1, 2, 3, 4, 5, and 6), and each one occurs with probability ⅙, the expected value (or weighted average) is

$$\tfrac{1}{6}(1) + \tfrac{1}{6}(2) + \tfrac{1}{6}(3) + \tfrac{1}{6}(4) + \tfrac{1}{6}(5) + \tfrac{1}{6}(6)$$

= 3.5. A reader may reasonably ask at this point how 3.5 can possibly be the expected value of this experiment, given that the only possible outcomes here are whole numbers. The point is that we do not expect 3.5 to ever actually occur as an outcome, but that, in the long run, it represents the *average* outcome. If, for example, somebody were to be paid one dollar for every dot shown on the die, we would expect the amount he was paid after 1,000 rolls of the die to be close to $3,500.

Here's another example, also involving the roll of a single die. Assume that a casino offers gamblers the following bet: the gambler puts down 1 dollar and is given a die to roll. If the gambler rolls a six, she wins three dollars (and gets her initial bet back). If she rolls a five, she gets her dollar back but does not win any additional money. If she rolls anything else, she loses her original dollar. Question: what is the expected value of this bet? Does it favor the gambler or the casino?

Sample Spaces and Events 13

To solve this, let us consider the outcomes from the gambler's point of view. There are three possible outcomes for her: she either makes three dollars of profit (+3), breaks even (0), or loses a dollar (−1). The probability that she wins 3 dollars is ⅙, as is the probability that she breaks even. The probability that she loses her dollar is ⅔. The expected value of her bet is, therefore,

$$\tfrac{1}{6}(3) + \tfrac{1}{6}(0) + \tfrac{2}{3}(-1)$$

which is −⅙. In other words, if the gambler plays this game a lot, she should expect to lose almost 17 cents for every dollar bet. The expected value *to the casino*, on the other hand, is ⅙ because whatever the gambler loses, the casino wins, and vice versa. Clearly, this game favors the casino.

One more example: most casinos offer a bet called "Big Red" (this is one of the side bets found at a craps table). In this bet, the gambler bets that the next roll of two dice will result in a sum of seven. If it does, the gambler wins $4 on a one-dollar bet (and gets her bet back for a profit of $4). If it does not, the bettor loses the initial one-dollar bet. To compute the expected value of this bet, note that (as we have previously observed) there are 36 possible outcomes when rolling two dice, six of which yield a sum of seven. So the probability that the gambler wins is ⅙ and therefore (complementary event!) the probability of losing is ⅚. This means that the bettor's expected value is ⅚(−1) + ⅙(4) = −⅙. As before, this means the bettor loses almost 17 cents per dollar bet. This is a terrible bet and should be avoided at all costs. By comparison, we will see in the next chapter that in roulette the expected long-term loss to the bettor is a little more than 5 cents per dollar. In other games like craps (Chapter 4), it is even smaller: less than a penny and a half per dollar. Indeed, very few bets in a casino offer a worse expected value for the bettor than does the Big Red bet.

Although our examples largely involve gambling, it should be noted that the concept of expected value applies in other "real life" situations. When an insurance company sets a premium for a life insurance policy, it is because the company has determined, from detailed actuarial tables, what the probability of the insured dying during the term of the policy is and made an expected value calculation as to what it should charge to make a specific profit. When an airline deliberately overbooks a flight, it does so knowing what the probability is of having to pay money if too many passengers show up, and therefore the expected value of risking this expenditure versus the expected value of passengers not showing up, and the airline making additional profit on the overbooked seats. (For an oversimplified illustration of this latter point, see the exercises.)

As should be clear from the previous examples, a negative expected value signifies a long-term loss, and a positive one signifies a long-term gain. (An expected value of 0 means the game is *fair*; in the long run, neither the casino nor the gambler loses.) No casino should (or knowingly would) offer a game that has a positive expected value for the gambler. However, even casinos can make mistakes. As recounted in [Bol4], in October 1994, the Grand Casino in Biloxi, Mississippi, offered the following bet: in exchange for a one-dollar bet, the gambler could roll three dice, and would win 80 dollars of profit if the sum of the three dice was 4; otherwise, the gambler lost his or her dollar. (The casino also offered the same terms on another bet, with 7 replacing 4.) We first note that the probability of rolling a sum of 4 on three dice is $3/256$; this is because the total number of possible rolls is, by the multiplication principle, $6 \times 6 \times 6 = 216$, and there are only three ways to roll a sum of 4: (1, 1, 2), (1, 2, 1), and (2, 1, 1). Now that we know the probability of the desired event, and therefore the probability of its complement, it is easy to compute the expected value, to the gambler, of this game: it turns out to be .125 (check this!). In other words, the gamblers *won* more than 12 cents per dollar, on average, playing this game. News of this payoff spread, and the casino became packed, resulting in a loss of about \$180,000 on this game in one day. The game was shut down for a day, and when it was reopened, the payoff was reduced from 80 to 60 dollars.

Exercises

1.30. Your friend offers you the following bet: you roll two fair dice. If you roll a sum of 2 or 3, you win three dollars; if you roll a sum of 12, you lose seven dollars; if you roll any other sum, you lose one dollar. What is the expected value to you of this game? Does it favor you or your friend?

1.31. Your friend offers you another bet: you roll a pair of fair dice three times. If you roll a sum of 2, 3, 11, or 12 in any of these three rolls, he pays you \$1. Otherwise, you pay him \$1. What is the expected value to you of this bet?

1.32. Your friend flips a dime and a quarter and gives you whichever coin or coins come up heads. If both coins come up tails, you give your friend 20 cents. Who, in the long run, will profit from this game? Explain.

1.33. Your friend now offers you another bet: you flip a fair coin. If it comes up tails, you pay your friend \$2. If it comes up heads, however, you flip it twice more, and your friend pays you one dollar for every head that appeared during the three flips. Determine all possible outcomes for you and the probability of each one, and use this information to determine the expected value of this game for you.

1.34. Verify the assertion, made in the last paragraph of this section, that the expected value of the game the casino offered is .125.

1.35. This question concerns a dice game called Pig. You play Pig according to the following rules: you get to roll either one or two dice (your choice). If any of the die comes up 1, you get nothing. If none of the dice comes up 1, you get the sum of the dice in dollars. So, there is some advantage to rolling two dice (you can potentially get a larger sum), but also some advantage to only rolling one die (smaller probability of getting a 1). In terms of expected value, which is the better strategy—roll one die or roll two?

1.36. In a certain National Park, a pass for the day for your vehicle costs $20. It is possible to enter the park without buying a pass (there are many roads into the park, and it is impossible to post an entry station on each one), but if you are caught in the park without a pass on your vehicle dashboard (which occurs, let us assume, with probability 1/10), you are fined $500. Obviously, the morally correct thing to do is buy a pass even if you think you won't be caught, but putting aside considerations of honesty and morality and thinking purely from a dollars-and-cents viewpoint, is buying a pass a wise decision? Explain your answer. (In other words, compare the expected value of a person buying a pass with the expected value of not buying one.)

1.37. An airline is considering whether to adopt a policy of deliberately "overbooking" a certain daily flight from Des Moines to Atlanta. The flight accommodates 20 passengers and is always full. Each ticket sold on the flight gives the airline a profit of 100 dollars. However, historical evidence gathered over a lengthy period of time shows that, on average, three out of ten times, a passenger will not show up for the flight even though he paid for a ticket. The airline is thinking that if they sold a 21st ticket for each flight, then three out of ten times they would be able to accommodate that extra passenger and make an extra 100 dollars for that flight. The drawback, of course, is that if all 20 ticketed passengers show up, which will happen on average on 7 out of every 10 flights, then the airline will not be able to accommodate the extra passenger and will have to make extra arrangements for him—buy him a ticket on another flight, put him up at a hotel, etc. The airline reasonably concludes that, in this case, they will incur a net loss of 150 dollars. If the airline adopts the policy of overbooking by one passenger per flight, what will they gain (or lose) on average per flight?

1.38. This question concerns the issue of whether a student should guess randomly on a certain standardized multiple-choice exam. The student encounters a question where he has no idea about the correct answer and is wondering whether to guess. The problem is that an incorrect answer is scored more harshly than a blank one: if a question is left blank, it scores 0 points, but if it is answered incorrectly, a deduction of $1/5$ point is made. Assume that the question has four possible answers, exactly one of which is correct. What is the expected number of points, on average, that a student will get per question if a random guess is made? What can you conclude about whether it is a good idea to guess in this situation?

1.39. You and a friend play a game in which you either win with probability p, or lose. If you lose, you pay your friend five dollars, and if you win, he pays you one dollar. Write down an expression, involving the unknown number p, for your expected value in this game. For what values of p is this game unfair to you? For what value of p is this game fair to both sides?

1.7 Odds

People in the gambling business often talk in terms of "odds" rather than "probability", but the terms are closely related: if we know the probability of an event, we can compute the odds of that event occurring, and vice versa. Although we will not use the "odds" terminology often in this book, we include a brief description of it for the sake of completeness.

When somebody says that the "odds of an event A happening are 3 to 2", that person means that the ratio of A occurring to A not occurring is 3 to 2. You can think of this as saying that for every three times A occurs, it does not occur (or A^C occurs) twice. In other words, in a five-element sample space, A occurs three times; so, it has probability $3/5$. If, instead, we said that the odds of A occurring are 4 to 1, then A would occur four times for every one occurrence of A^C, and this would mean that the probability of A is $4/5$.

The statement that the odds *against* A are 3 to 2 means the same as the statement that the odds in favor of A^C are 3 to 2. So, if the odds against A are 4 to 1, then that means that A^C has probability $4/5$, or that A has probability $1/5$. Another way to think about this is to say that if the odds against A are "p to q" then the odds in favor of A are "q to p".

It is easy to go in the reverse direction and, given the probability of an event, determine the odds of it happening. For example, if the probability of an event is $3/8$, then that means that it happens 3 times out of 8, or that the event happens three times for every five that it does not, i.e., that the odds of the event are 3 to 5. In this case, the odds against the event would be 5 to 3.

In a situation like the toss of a fair coin, where heads and tails are equally likely, then the event "heads" occurs (in the long run) once for every time "tails" does, so the odds of "heads" occurring are 1 to 1.

The word "odds" gets used in another way in gambling as well. The phrase "house odds" refers to what the casino pays off on a successful (to the gambler) bet. As an example, if the house odds on a certain bet are 35 to 1, that means the gambler wins 35 dollars for every dollar bet. (This is, in fact, the payoff on a single-number bet in roulette, a game that we will discuss in more detail in the next chapter.) To distinguish "house odds" from the kind of odds that we have been discussing in this section, we might refer to the latter as "mathematical odds" or "true odds".

Sample Spaces and Events

Exercises

1.40 If the odds of an event occurring are 4:3, what is the probability that the event occurs? What are the odds against the event occurring? What is the probability that the event does not occur?

1.41 If an event has probability 3/8, what is the probability that the event does not occur? What are the odds of the event occurring? What are the odds against the event occurring?

1.42 The odds in favor of an event A are 3:2, and the odds in favor of an event B are 2:1. Which event is more likely to occur? Explain.

1.43 The odds in favor of an event A are 1:4. The house odds for this event are 3:1. What is the expected value, to the gambler, of this event?

1.44 In the previous question, what would the house odds have to be in order for this bet to be fair?

2
Roulette

Even the relatively small amount of probability theory that we discussed in Chapter 1 is sufficient to allow us to draw some interesting mathematical conclusions about roulette, a mainstay of casinos in the United States and Europe and one of the most famous of all casino games. In this chapter, we will look into some of the standard wagers in this game and explore what the casino's expected profits on these wagers are. We will close the chapter by discussing whether there is such a thing as a roulette "system".

2.1 Rules of the Game

As many of the readers of this book probably already know, the game of roulette is played with a wheel that is divided into 38 slots. In the United States (we will discuss European wheels shortly), two of these slots are denoted as 0 and 00, respectively, and are colored green. The remaining slots are denoted as 1–36; half of these slots are colored black, and the remaining ones are colored red (Figure 2.1). The slots are not consecutively numbered on the wheel, but the colors red and black do alternate. Consecutively numbered slots do not necessarily have different colors; for example, the slots numbered 18 and 19 are both red.

To play the game, the wheel is spun, and a little ball is released. The ball spins around the wheel in the opposite direction until it eventually slows down and falls into one of the pockets. The gamblers at the roulette table have placed bets based on color and/or number; for example, one might bet "even", in which case that person wins if the ball lands in one of the even slots from 1 to 36. (The number 0, though considered an even number in mathematics, is neither even nor odd for purposes of roulette; neither is 00.) A winning even/odd or black/red bet pays even-money—i.e., the gambler wins in profit the amount of his or her bet.

In addition to the four even-money bets specified above, there is also a "single number" bet, in which the gambler bets on a single number. A winning single number bet pays a profit of 35 times the amount of the bet. For example, a person who has bet 5 dollars on number 6 will make a $175 profit if the ball lands in that pocket.

Roulette

FIGURE 2.1
Roulette Wheel.

There are other bets as well. In general, one places a bet in roulette by putting a chip worth a certain amount of money down on the roulette playing table near the wheel. (See Figure 2.2 for a picture of the playing table.) So, for example, one can place a bet on one of the three columns of numbers shown (making a profit of twice your bet if any of the 12 numbers in the column show up).

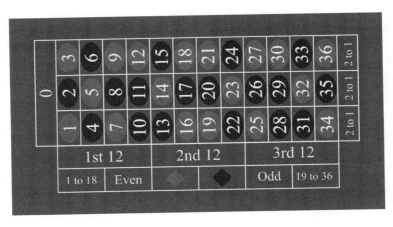

FIGURE 2.2
Roulette Table Layout.

Roulette wheels in Europe are a little different. Instead of having both a 0 and a 00, a European wheel has only a 0. So, instead of having 38 slots, a European wheel has 37. The remaining 36 numbers are, as in an American wheel, numbered 1–36, with these slots being divided equally between red and black.

Exercises

2.1. Suppose that, on an American wheel, John places a $1 bet on the number 15 and another $1 bet on the number 20. What are the possible outcomes for John? (Consider various possibilities depending on what number shows up. For example, if the number 0 shows up, John loses both bets, for an outcome of −2.)

2.2. Suppose that, on an American wheel, John places a $1 bet on the number 17 (which is black) and another $1 bet on the color black. What are the possible outcomes for John?

2.3. Suppose that, on an American wheel, John places a $1 bet on the color red and another $1 bet on the color black. What are the possible outcomes for John?

2.2 Some Roulette Calculations

Now that we understand the mechanics of the game, let us use mathematics to analyze it. Of course, on any given spin of the wheel, the casino will win some of the bets that have been made and lose some others. However, as we have seen, what the casino is really interested in is their long-term average profit or loss—i.e., the *expected value* of the bet. Recall that this number is calculated by taking every possible outcome of a bet, multiplying that outcome by the probability that it occurs, and then adding up these products. In other words, we are taking a "weighted average" of these outcomes. In doing the calculations that follow, we assume for simplicity that the gambler bets $1, even though in many casinos a bet this small would not be allowed.

Suppose first that the gambler places a $1 "red" bet. We will compute the gambler's expected value of this bet. The gambler will either win and make a $1 profit (outcome: +1) or lose his initial bet (outcome: −1). The probability that the gambler wins is very easy to compute: there are 38 possible pockets for the ball to go into, and 18 of these are colored red, so the probability of a win is $18/38$. Likewise, the probability of a loss is $20/38$. (This is the complimentary event; alternatively, you can note that there are 20 ways to lose: 18 black numbers and 2 green ones.) So, the expected value to the gambler of this bet is (multiply each outcome by its probability and add together) $-2/38 = -1/19$, which is approximately −.0526. In other words, a gambler playing a red bet repeatedly will, in the long run, lose a little more than a nickel for every dollar bet. Since the casino wins what the gambler loses, it will, in the long run, win a little more than a nickel for every dollar bet. Of course, the exact same calculations show that this is also the expected value for a black bet, an even bet, or an odd bet. Think about how many times one of these bets is made in a casino, and think of the casino, in the long run, collecting more than five cents per dollar on each of these bets, and you begin to see why casinos generally do quite well.

Recall from the previous section that European wheels lack a 00 slot. What effect does the removal of the 00 have on an even-money bet like red? Now, there are 37 slots, 18 of them red, so the probability of winning is $18/37$ instead of $18/38$. If we mimic the computation that we used in the last paragraph, we see that the gambler's expected value is now $-1/37$, which is approximately −.027. The "house advantage", therefore, has been roughly cut in half by the removal of the 00 slot.

Now consider a single number bet. The probability of winning here is $1/38$ and you make a profit of 35 dollars on a one-dollar bet. The expected value of this bet to the gambler is therefore

$$35\left(1/38\right) - 1\left(37/38\right) = -2/38 = -1/19,$$

which is precisely the expected value of an even-money bet. So, even though a single number bet certainly seems like a much riskier bet, it turns out that, over time, the average loss to the gambler is no worse than it is for an even-money bet. In fact, the interesting thing about roulette is that all the standard bets have the same expected value: the casino wins, on average, a little more than five cents for every dollar bet.

On a European wheel, the gambler's expected value of a one-dollar single number bet is, as a simple calculation shows, $-1/37$, the same as an even-money bet.

Returning to America, let us see what happens if a gambler combines bets. For example, suppose the gambler puts $1 on "red" and $1 on the single number 20 (which is black). What is the expected value of this bet? As always, we begin to answer this question by figuring out the possible outcomes for the gambler. It's not possible to win both bets because, by definition, if one occurs, the other does not. So, the best possible outcome for the gambler would be to win the bet that pays the most, the number 20 bet. In that case, he or she makes a 35-dollar profit on that bet but loses one dollar on the red bet; the net outcome is therefore a 34-dollar profit. The probability that this will happen is $1/38$. The next best outcome for the gambler would be to win the red bet; in this case, he or she makes a $1 profit on this bet but loses a dollar on the red bet, so the net outcome for the gambler is 0. Finally, the worst that can happen for the gambler would be to lose bets (net outcome −2); for this to happen, the wheel must come up with a non-red number other than 20. There are 20 non-red numbers (18 black numbers plus the 0 and 00), one of which is 20, so there are

19 ways to lose both bets. The probability of this happening, of course, is $19/38$. The expected value of the gambler's double bet is $34(1/18) + 0(18/38) - 2(19/38) = -4/38$. This is twice the expected value of the other bets we have looked at, but remember that the gambler is betting $2 on this bet, not $1. To arrive at the expected value *per dollar*, we must divide by 2, and we see that, here too, this bet results in a loss of a little more than five cents per dollar.

Another mathematical aspect of roulette play should be noted here. Many casinos have an electronic screen above a wheel, recording the recent results on that wheel. A number of people use these results to assist them in making a bet. For example, suppose the last five rolls have all shown red. Is black more likely to come up on the next roll? Quite a few people instinctively think that the answer is yes—that a black number is "due". But in fact, this is simply wrong—the probability of a black number is $18/38$, a calculation that does not depend on what happened previously. (To put it another way, the wheel has no memory. In mathematical terms, the spins of a roulette wheel are *independent* events, a concept that will be discussed in more depth in the next chapter.) So, any kind of betting strategy that is based on the results of previous spins of the wheel is flawed from the outset. Indeed, according to [Bol4], these electronic signs are money-makers for the casinos, and according to one casino, a sign that the casino installed paid for itself within about six weeks.

Exercises

2.4. Consider the bet described in exercise 2.1. What is the expected value to the gambler of this bet?

2.5. Consider the bet described in exercise 2.2. What is the expected value to the gambler of this bet?

2.6. Consider the bet described in exercise 2.3. What is the expected value to the gambler of this bet?

2.7. Verify the statement, made in the text, that on a European wheel, the gambler's expected value of a one-dollar single number bet is $-1/37$.

2.8. In some European casinos, a player who places an even-money bet will, if a 0 comes up, get half his or her wager back instead of losing the entire thing. What is the expected value of a $1 bet if this rule is followed?

2.9. A certain roulette wheel is weighted so that the probability of getting "red" is $15/38$ instead of the customary $18/38$. What is the expected value to the bettor of a red bet?

2.10. Go to an online roulette site (such as https://www.247roulette.org/) where you can spin a roulette wheel for free without downloading anything and play 20 rounds of an even-money $1 bet. According to the calculations done in this chapter, you should wind up losing about one dollar. Do your results conform to this?

2.3 Roulette "Systems"

It's not too hard to find on the Internet a number of websites discussing (or offering to sell) roulette "systems"; i.e., methods of play that are supposed to produce favorable outcomes. In this section, we will prove that no such system can exist (at least, no honest system can). We will then discuss one such purported system that actually sounds superficially plausible; we will then explain why, however, it does not work.

Actually, we have *already* proved that no such system can exist. We know that in the long run, the casino will make more than a five cent profit on every dollar bet in roulette. This is true for every dollar bet, regardless of how the bettor came to make the bet. So, it follows that no honest system can ever produce a winning strategy—while a gambler may get lucky and win for a while, he or she cannot, in the long run, beat the odds.

So, if you see a website offering to sell a foolproof system for betting at roulette, you should be aware that the only person who will assuredly benefit from that sale will be the person selling the so-called "system".

This is not to say, however, that some systems may sound pretty plausible, at least upon superficial examination. We will discuss one such system in the remainder of this section, called the Martingale system.

This system is based on a simple mathematical formula, which we discuss below. First, let us note the following equations, all of which can be mentally verified:

$$
\begin{aligned}
1+2 &= 4-1 \\
1+2+4 &= 8-1 \\
1+2+4+8 &= 16-1
\end{aligned}
$$

Because $2 = 2^1$, $4 = 2^2$, $8 = 2^3$, $16 = 2^4$, these equations can be seen as special cases of the equation

$$1 + 2 + 4 + \cdots + 2^n = 2^{n+1} - 1 \quad (*)$$

Equation (*) is itself a special case of the formula for the sum of a geometric progression, which the reader may have encountered in high school. Proving it carefully for all positive integers n requires the Principle of Mathematical Induction and would take us too far afield. We will simply take it as given.

With equation (*) in hand, we now describe the Martingale system. For the sake of simplicity, we assume that the gambler starts with a $1 bet, even though casino rules may require a minimum bet in excess of that amount. You can consider a "one unit" bet instead of a $1 one if you like, where "one unit" might be, for example, a 5-dollar chip. Here are the rules for the Martingale system:

- Begin by placing a $1 even-money bet, say a bet on black.
- If you win that bet, stop and walk away; you have made a profit. If you lose, double that amount and make a $2 black bet.

- If you win the $2 bet, stop; you win $2, and, even considering the $1 you lost in the previous bet, you walk away a $1 winner. If you lose this bet too, double your bet and make a $4 black bet.
- If you win the $4 bet, stop and walk away. You are again a winner; you have already lost 3 dollars, but by winning 4, you are again one dollar ahead.
- If you lose the $4 bet, double it and make an $8 bet. If you win this, you win $8. Since you have lost $1 + 2 + 4 = 7$ dollars, you again come out ahead.
- Keep this up until you eventually win, which, since the probability of doing so is almost ½, should occur fairly soon. At the point that you have won, equation (*) says that the amount you win (a power of 2) is one more than the sum of all smaller powers of 2, so you emerge from the game as a winner. True, you've only won $1, but you have won.

This does seem like a reasonably plausible system since most people think it is very unlikely that a bet like black, which has a probability of almost ½ of occurring, will not happen in three or four spins of the wheel. So, where is the flaw in this system?

For one thing, recall that individual spins of a wheel do not depend on previous spins, so there is no guarantee that a black number will show up in just a few spins. In fact, the Internet tells us that in 1943, the color red showed up at a casino 32 times in a row on one roulette wheel. This is apparently the largest number of consecutive spins of the same color in an American casino. Think about what would happen if you were playing at an American casino using the Martingale system—your next bet would be for $2^{32} = 4,294,967,296$ dollars. In short, you would be betting more than 4 million dollars, all in the hope of walking away with a $1 profit.

It's a pretty safe bet that most people don't have this kind of money to throw away on a single bet at roulette, and even if they did, the casino probably wouldn't let them do it—most casinos have maximum bet rules in place. The Martingale system can, therefore, easily wind up bankrupting the gambler if there is a long string of losing spins.

So, to summarize: in the short run, this system may produce a small profit, but in the long run, you can't escape the mathematics: you will lose more than five cents for every dollar played.

This concludes our brief introduction to roulette. There are any number of websites available where a person can play roulette online without having to download anything first (one good site, at least as of this writing, is https://www.247roulette.org/), and the reader may enjoy finding one such website and playing a number of games, checking his or her outcome with that predicted by the expected value of the game. (See exercise 2.9.)

Exercises

2.11. John decides to use the Martingale system, making successive "red" bets, starting with a $1 bet. Suppose he loses the first five bets. What must he bet on the sixth turn of the wheel? Would you feel comfortable betting this amount just to wind up with a $1 profit? What would John's bet have to be if his initial bet was $5?

2.12. With reference to the preceding problem, what is the probability that John will, in fact, lose his first five bets in a row?

2.13. Go to the website cited above and play the Martingale system several times. How many spins of the wheel did it take before you showed a profit?

3

Conditional Probability and Independence

The mathematical concept of conditional probability simply expresses the common-sense idea that knowledge of a situation may alter the probability of an event. For example, suppose you select the top card from an ordinary, well-shuffled deck, and let A denote the event "you draw the ace of spades". We already know how to compute the probability of A: it is clearly $1/52$, since there are 52 possible outcomes, only one of which is the ace of spades. But suppose now that when the deck was being shuffled, you observed that the two of diamonds landed on the bottom. This knowledge changes the probability of A because there are now only 51 possible outcomes; we know the top card is not the two of diamonds. So, in mathematical terms, the probability of A, given that the top card is not the two of diamonds, is $1/51$. This is an example of a conditional probability problem.

3.1 Definition of Conditional Probability

In general, if A and B are events defined on the same sample space, the "conditional probability of A, given B", is the probability of A, given that the event B has occurred. This is denoted as $P(A|B)$; it is read as "the probability of A, given B". For example, in the illustration given in the preceding paragraph, A could be the event "you draw the ace of spades" and B the event "you do not draw the two of diamonds".

There is a formula for $P(A|B)$; this is mathematics, after all, so there is always a formula. But it seems more insightful and intuitive to look at another way of computing a conditional probability that does not involve just plugging numbers into an unmotivated formula. Recall that what really happened in the example above was that the sample space changed once we assumed the conditional event. That's the key to computing conditional probabilities: we need to adjust our sample space to create one that is consistent with the event B occurring.

For example: Suppose we roll a single die. Let A be the event "the number shown is 2 or 4" and B the event "the number shown is even". If you were simply asked to compute $P(A)$, the answer would be $2/6$ $(= 1/3)$, since in this computation there would be six elements in the sample space, and two of them (2 and 4) correspond to a favorable outcome for A. But if we are asked to compute P(A|B), we are by definition assuming the die came up even (that's what "given B" means), and therefore we must rule out, at the outset, the possibility that the die came up 1, 3, or 5. In other words, our adjusted sample space now consists of the possible even outcomes, namely 2, 4, and 6. Since there are three possible outcomes now, two of which are favorable (i.e., result in event A occurring) then the conditional probability is ⅔.

For the events A and B described in the preceding paragraph, we can also compute P(B|A), the conditional probability of B given A. Our new adjusted sample space must be the subset of the original one consisting of outcomes that are consistent with A. There are only two of them: 2 and 4. Both of these numbers, of course, are even, so event B is guaranteed to occur if we assume event A. Thus, P(B|A) = 1.

Before looking at more examples, let us summarize the steps involved in computing P(A|B) for events A and B:

- Step 1: Adjust the sample space by eliminating any outcomes inconsistent with B.
- Step 2: From the adjusted sample space, count the number of favorable outcomes (i.e., outcomes that are consistent with A).
- Step 3: Divide the number from Step 2 by the number from Step 1.

We close this section with a few other examples.

> **Example 3.1.1.** A coin is flipped three times. What is the probability that there are exactly three heads, given that there are at least two heads?
>
> **Solution.** From our original eight-element sample space, we remove the outcomes inconsistent with there being at least two heads, giving us, as an adjusted sample space, the set {HHT, HTH, THH, HHH}. Of the four elements in this set, there is only one consistent with the event "exactly three heads", namely HHH. So, our desired conditional probability is ¼.

The next example is a famous one because it is very counter-intuitive. If your intuition initially leads you to the wrong answer, don't feel bad; that happens to many people.

> **Example 3.1.2.** You meet Mr. Smith at a party. He tells you that he has two children, ages 5 and 8. He also tells you that at least one of his children is a girl. What is the probability that they are both girls? (You may assume that the sex of one child is independent of the sex of the other and that a child is equally likely to be a boy or girl.)
>
> **Solution.** Most people say the probability is ½ because any child is equally likely to be a boy or girl. This is incorrect, however, and here is why: we can think of this problem as asking for P(C|D), where C is the event "both girls" and D is the event "at least one girl". To find this

conditional probability, we first consider sample spaces. The sample space for the sex of two children, under the simplifying assumptions made previously, is clearly {BB, BG, GG, GB}, where the first entry in every one of these four elements is the sex of the younger child, the second is the sex of the older one, and B and G denote, of course, "boy" and "girl". To adjust this sample space to accommodate event D ("at least one girl"), we must eliminate BB, thereby giving us a three-element sample space. Of these three elements, there is only one that is consistent with event C (namely GG), so the conditional probability that we seek is ⅓ rather than ½.

Why is the intuitive answer ½ incorrect? While it is true that the probability that one specific child is a girl is ½, we are not dealing with a specific child here: Mr. Smith did not give you enough information to allow you to focus on a particular child. Notice how important language is in this example. Had Mr. Smith told you, "my older child is a girl", then the probability that both children were girls would, in fact, be ½. This is because our adjusted sample space in this case would now consist of two elements, namely BG and GG; the extra information supplied by Mr. Smith would allow you to rule out the possibility of GB as well as BB.

Here is another interesting example, one that will recur later when we study the game of craps.

Example 3.1.3. Jane rolls two dice and records their sum. She continues to do this until she obtains a sum of 4 or a sum of 7. What is the probability that she rolls a sum of 4 before a sum of 7?

Solution. At first glance, this doesn't even appear to be a conditional probability problem at all. However, a simple trick (looking at this problem in a different light) allows us to recast it as one. First, notice that any roll that does not produce a sum of 4 or 7 is simply irrelevant; it adds nothing to the discussion and can be ignored. Thus, the only kind of roll that is of interest to us is one where the sum is either 4 or 7. It therefore follows that the problem may be rephrased as: what is the probability that the sum of a roll is 4, given that it is either 4 or 7? Once rephrased, the problem becomes easy to solve. We have already seen that the sample space for a roll of two dice has 36 elements. If we adjust this sample space by removing entries that reflect a sum other than 4 or 7, we are left with nine elements: (1, 3), (2, 2), (3, 1), (1, 6), (2, 5), (3, 4), (4, 3), (5, 2), and (6, 1). Of these nine elements, three of them are favorable. Therefore, the desired answer is ⅓.

As noted earlier, there is also a formula for P(A|B). To derive the formula, note that the adjusted sample space can be thought of as a sample space for the event B, and then we consider the elements in that space that are consistent with A as well as with B—i.e., the outcomes that correspond to the event AB (A and B). So, our formula for P(A|B) is

$$P(A|B) = P(AB)/P(B).$$

Conditional Probability and Independence

Actually thinking about what the adjusted sample space is, however, and proceeding as in the examples above, still seems like the more intuitive way of solving a conditional probability problem, at least if the sample spaces involved are manageable in size.

We end this section with a cautionary note: in dealing with conditional probabilities in real life, care must be taken with how they are handled. One example of this is the so-called "Prosecutor's Fallacy", which we discuss briefly below. A full discussion, involving such things as Bayes' Theorem, would be rather involved and subtle, so we will omit it.

To set the stage for the Prosecutor's Fallacy, suppose that a murder has been committed. Forensic examination of the murder scene shows that the murderer has a very rare genetic marker in his or her blood. John happens to have this marker and is arrested for the murder. The prosecutor, lacking any eyewitness testimony or evidence of motive, makes a statistical argument to the jury: because this marker is so rare, the probability that any innocent person would have it is therefore microscopically small, and so John must be guilty.

This argument does have a certain superficial appeal (and variations of this argument have, indeed, been made in courtrooms before). However, let's think about what is really going on here. The prosecutor has just pointed out the conditional probability

$$P(\text{Person has the genetic marker} \mid \text{Person is innocent})$$

is very small. Well, of course it is; the probability that *any* person, guilty or not, would have it is small. This, however, is not the probability that the jury should be interested in. In view of the fact that John *has* the marker, the critical question for the jury is the value of

$$P(\text{Person is innocent} \mid \text{Person has the genetic marker})$$

which, we have seen previously, is not necessarily the same number.

We will briefly return to the Prosecutor's Fallacy in Section 3.3. Those readers who have an interest in mathematics and the law might enjoy the book [SC], which is filled with interesting examples of how mathematics has been misused in the courtroom.

Exercises

3.1. If two dice are rolled, what is the probability that their sum is 8, given that one of the die shows a 3?

3.2. In the previous problem, what is the probability that one of the die shows a 3, given that the sum is 8?

3.3. Two dice are rolled until a sum of 8 appears. What is the probability that when this happens, each die will show a 4?

3.4. Two dice are rolled. What is the probability that both dice show an even number, given that the sum is 10?

3.5. Two dice are rolled. What is the probability that the sum is 10, given that both dice show an even number?

3.6. A fair coin is tossed five times. What is the probability that all five tosses resulted in heads, given that there are at least four heads?

3.7. A fair coin is tossed five times. What is the probability that all five tosses resulted in heads, given that there is at least one streak of three heads in a row?

3.8. A fair coin is tossed five times. What is the probability that there are four consecutive heads, given that there are at least four heads?

3.9. The "Ask Marilyn" column in the July 31, 2011, issue of Parade Magazine contains the following question from a reader: "Say you plan to roll a die 20 times. Which of these results is more likely: (a) 11111111111111111111, or (b) 66234441536125563152?" Marilyn's response, reprinted verbatim, was: "In theory, the results are equally likely. Both specify the number that must appear each time the die is rolled. (For example, the 10th number in the first series must be a 1. The 10th number in the second series must be a 3.) Each number—1 through 6—has the same chance of landing face-up. But let's say you tossed a die out of my view and then said that the results were one of the above. Which series is more likely to be the one you throw? Because the roll has already occurred, the answer is (b). It's far more likely that the roll produced a mixed bunch of numbers than a series of 1's." Is Marilyn correct? Explain.

3.10. At a certain small university, there are 1,000 male students and 800 female students. Of the 1,000 male students, 400 are entered in the Bachelor of Arts (BA) program, and the other 600 are entered in the Bachelor of Science (BS) program. Of the 800 female students, 700 are entered in the BA program, and the remaining 100 are entered in the BS program. One student is selected at random from the student population. What is the probability that the student is female, given that he or she is in the BA program?

3.11. (See previous question.) What is the probability that the student is in the BA program, given that she is female?

3.2 Law of Total Probability

We begin this section with a question. There are two urns, one containing a single white ball and the other containing three black balls. Jane flips a fair coin to determine which urn she selects a ball from. If the coin is heads, she selects the first urn; otherwise, she selects the second. What is the probability that Jane ultimately selects a white ball?

Conditional Probability and Independence

Some students have been known to say ¼, because there are a total of four balls, one of which is white. But this is incorrect because Jane never winds up drawing one ball from four. She either draws a ball from the first urn, in which case she is selecting one ball out of 1 (and therefore draws a white ball with probability 1), or she draws a ball from the second urn, in which case she selects one ball out of 3. But in this case, too, the outcome is preordained: she must draw a black ball. Therefore, if she flips heads, she draws a white ball; if she flips tails, she draws a black ball. The probability of drawing a ball of either color is therefore ½.

We can think of this as a computation involving conditional probabilities, as follows: let B and W be the events "she draws a black ball" and "she draws a white ball", respectively. Let U_1 and U_2 be the events "she draws from urn 1" and "she draws from urn 2", respectively. Then we have

$$P(W) = P(W|U_1)P(U_1) + P(W|U_2)P(U_2) = \tfrac{1}{2}, \text{ since}$$

$P(W) = \tfrac{1}{2}$, $P(W|U_1) = 1$, $P(U_1) = \tfrac{1}{2}$, $P(W|U_2) = 0$, and $P(U_2) = \tfrac{1}{2}$. This formula is an illustration of the *Law of Total Probability*, which we now state.

Law of Total Probability: Let A be an event. Suppose also that B_1 and B_2 are two events, one of which must occur, but which cannot both occur. Then

$$P(A) = P(A|B_1)P(B_1) + P(A|B_2)P(B_2)$$

This law extends naturally to more than two events: B_1 and B_2. The point is that we have a finite set $B_1, \ldots B_n$, one of which must occur, but which are exclusive in the sense that no two of them can occur at the same time. Then, for an event A, P(A) is the sum of all the terms $P(A|B_i)P(B_i)$.

Example 3.2.1. Officials at a large state university want to take a poll among students to find out what percentage of them have ever cheated. In other words, the officials know that a certain percentage, p, of students cheat, and would like to try and estimate p. The pollsters know that if they just ask each student, "have you ever cheated on a test?" Most people, either out of shame or fear of retribution, will say "no" even if they have, in fact, done so. So, the pollsters decide to be clever, and they hand each person a die and say, "Roll the die, but don't let me see the result. If it comes up 1 through 4, answer the question honestly. If it comes up 5 or 6, answer the question dishonestly." Under these circumstances, a "yes" answer to the question is no longer stigmatizing or incriminating because the student may very well, as per instructions, be answering the question dishonestly; hence, the pollsters believe they can expect valid answers now. Assuming that ⅖ of the students answer "yes" to this question, what does that say about the value of p?

Solution. Let Y be the event "a student answers yes", B_1 the event "the student rolled a 1, 2, 3 or 4", and B_2 the event "the student rolled a 5 or 6". By the Law of Total Probability, we have

$P(Y) = P(Y|B_1)P(B_1) + P(Y|B_2)P(B_2)$. We can assume that $P(Y)$ is roughly $2/5$, since that is the proportion of students who answered "yes". What is $P(Y|B_1)$? A student answers "yes" honestly if and only if that student cheated, so this probability is approximated by the number p. Likewise, $P(Y|B_1)$ is roughly $1 - p$, since a student will answer "yes" dishonestly if and only if that student has never cheated. It is clear that $P(B_1) = 2/3$ and that $P(B_2) = 1/3$, since these are the respective probabilities that a student rolls a number between 1 and 4, and that a student rolls a 5 or 6. Plugging all these numbers into the equation $P(Y) = P(Y|B_1)P(B_1) + P(Y|B_2)P(B_2)$ gives (check this!) an equation in p, the solution to which is $1/5$.

Here is one final example.

Example 3.2.2. John's father offers him an unusual birthday present. John flips a coin. If it comes up heads, his father gives him five dollars. If it comes up tails, he then rolls a fair die and receives, in dollars, whatever number the die shows. What is the probability that he receives five dollars for his birthday?

Solution. Let F be the event "John gets five dollars". Let H and T be the events "he rolls heads" and "he rolls tails", respectively. Since H and T cannot both occur, but at least one of them must occur, we can apply the Law of Total Probability, which in this notation reads $P(F) = P(F|H)P(H) + P(F|T)P(T)$. Observe that $P(H) = P(T) = 1/2$, $P(F|H) = 1$, and $P(F|T) = 1/6$. Plugging in these values and computing, we see that $P(F) = 7/12$.

Exercises

3.12. A box contains four black balls and three white balls. John removes one ball and puts it aside without looking at it. He then removes a second ball. What is the probability that the second ball chosen is black? Do this first by using the Law of Total Probability, but then explain why that really wasn't necessary.

3.13. Box 1 contains four black balls, two green balls, and a white ball. Box 2 contains five green balls. John flips a coin to determine which box he will remove a ball from, and then he removes a ball from that box. What is the probability that the ball is white? What is the probability that the ball is green? What is the probability that the ball is black?

3.14. John rolls a single die. If it comes up with one of the numbers 1, 2, 3, or 4, John is given ten dollars. If the die comes up 5 or 6, however, John is then told to flip a coin. If it comes up heads, he is given ten dollars. If it comes up tails, he is given five dollars. What is the probability that John wins ten dollars? What is the probability that John wins five dollars?

Conditional Probability and Independence

3.15. (Continuation of preceding exercise.) If John plays this game over and over again, what, on average, will he win per game?

3.16. John plays a game. He first flips a coin. If it comes up heads, he wins four dollars. If it comes up tails, he rolls a die and wins the amount shown (in dollars). What are John's expected winnings per game?

3.17. In the campustown section of a certain college town, there are three bars (A, B, and C) lined up in a row, in that order, on a certain street. At 7 PM on a Friday night, a student enters one of the bars at random (with equal probability of selecting any of the bars) and stays there for one hour, drinking steadily. At 8 PM that student then leaves that bar to go to an adjacent one. If the student was already in one of the bars at the end (A or C), then there is only one adjacent bar, namely B, and the student will definitely go to it; if the student was in bar B, he will pick one of the two adjacent bars at random. What is the probability that, at 8 PM, the student goes into Bar A? Bar B? Bar C?

3.3 Independent Events

We know from prior work that if A and B are nonoverlapping (i.e., disjoint) events in a sample space S, then the event "A or B" (which as a set is the union $A \cup B$) has probability $P(A \cup B) = P(A) + P(B)$. It seems natural to ask whether there is a corresponding formula for the event "A and B", given by the intersection $A \cap B$ or AB. A natural guess might be $P(AB) = P(A) P(B)$.

How might we investigate the truth or falsity of this formula? One obvious approach would be to look at examples and see if the formula works in these cases. As a first example, suppose we draw a card from an ordinary deck, and let A and B be the events "a Queen is drawn" and "a heart is drawn", respectively. In this case, it is easy to see (check this!) that $P(A) = 1/13$, $P(B) = 1/4$, and $P(AB) = 1/52$. So, in this case, the formula does work, leading us to perhaps hope that we have discovered a correct formula.

But now let us consider a different example, this time also involving a card drawn from a fair deck. Let A be the event "a red card is drawn" and B be the event "a black card is drawn". Now we have $P(A) = P(B) = 1/2$, but $P(AB) = 0$. So, the formula does not, in this case, work. It is, therefore, *not* a universal rule that the product of the events P(A) and P(B) is P(AB).

Can we, however, perhaps discern something useful from a study of these examples? Notice that in the second example, if we know that event A has taken place, then we also know that event B has not. In other words, knowledge of event A's occurrence tells us something (a great deal, actually) about event B's occurrence. But in the first example above, this is not the case. Since hearts are spread uniformly throughout the deck (one for each rank), knowing that a Queen has been drawn tells us nothing about whether a heart has been drawn.

This turns out to be the key observation in determining whether P(AB) = P(A) P(B). Perhaps this is not so surprising if we remember the formula for conditional probability that we ended Section 3.1 with P(A|B) = P(AB)/P(B). If we replace P(AB) in this formula with P(A) P(B), the right-hand side becomes P(A). Thus, if the multiplication formula works, then P(A|B) = P(A), or, putting it another way, the fact that B occurs does not change the probability of A. In other words, A and B are *independent* events.

In this book, we will think of independence in an intuitive way, asking whether one event affects the other. But in more advanced books on probability, it is the formula P(AB) = P(A) P(B) that is used to *define* when two events are independent.

Example 3.3.1. A fair die is thrown. Let A be the event {1, 2} and B the event {2, 4, 6}. Are A and B independent?

Solution. A simple calculation shows that P(A) = ⅓, P(B) = ½, and P(AB) = ⅙. Because P(AB) = P(A) P(B), these events are independent.

Example 3.3.2. A fair die is thrown. Let A be the event {2} and B the event {2, 4, 6}. Are A and B independent?

Solution. This time, because P(A) is now ⅙, the formula is not satisfied, and the events are not independent. It is worth looking at these two examples intuitively. If we know that either 1 or 2 has been rolled, then we don't know whether 2, 4, or 6 have. But if we know that 2 has been rolled, then we certainly know that event B has occurred.

Example 3.3.3. A fair coin is tossed three times. Let A be the event "at least one toss is heads" and B the event "at least one toss is tails". Are A and B independent?

Solution. Intuition tells us the events are not independent, because if we know at least one toss is a head, then it is less likely that one is a tail since there are fewer opportunities for a tail to show up. Let us look at this mathematically. The sample space for this experiment has, as we have seen, eight elements in it, seven of which are favorable to both events A and B. So, P(A) = P(B) = ⅞. On the other hand, for the event AB (both A and B), six elements of the sample space are favorable, so P(AB) = ¾. Since P(AB) is not equal to P(A)P(B), the events are not independent.

In Section 3.1, we pointed out that careless use of conditional probability in such venues as, say, courtrooms, could cause serious real-life consequences. The same is true of careless use of arguments involving independent events. To illustrate, we look at the case of Sally Clark, a British solicitor who, in the late 1990s, was convicted of the murder of her two infant sons.

Clark's first son died in December 1996, an apparent victim of "cot disease", what we in the United States call Sudden Infant Death Syndrome (SIDS). A little more than a year later, Clark's second son died under similar circumstances, and Clark was arrested for the murder of both infants.

Conditional Probability and Independence

There was no evidence that Clark had any motive to kill her sons; the prosecution's case was largely based on statistics. These statistics came from the testimony of Roy Meadow (a pediatrician, not a mathematician or statistician); he testified that in an affluent family like the Clark's, only about 1 in 8,500 infants would be expected to die of SIDS. Meadow thus concluded that the probability of two children dying of SIDS would be $(1/8500)^2$, which is roughly 1 in 73 million. Clark was convicted on the basis of the fact that this number is so small.

Hopefully, the reader sees the problem with this reasoning: multiplying the probabilities is legitimate only if we know that the event of one child dying of SIDS is independent of the other child also dying of that disease. But of course, our intuition tells us precisely otherwise: SIDS may well be the result of some genetic factor in the family.

Clark was eventually released from prison, but not because of the mathematically illegitimate evidence that led to her conviction. Instead, evidence came to light that the prosecution's pathologist was aware of exculpatory evidence on Clark's behalf but did not share this information with the defense. Clark served more than three years of her sentence and came out of prison a broken woman. She died in 2007 of alcohol-related issues.

In 2001, the British Royal Statistical Society issued a news release condemning the shoddy evidence used to convict Clark and warning against similar mistakes in the future.

A similar misuse of probability involving non-independent events occurred here in the United States, specifically in California in 1968. This was the case of *People v Collins*, which is famous enough to have its own Wikipedia page: https://en.wikipedia.org/wiki/People_v._Collins, which we encourage the reader to look at. It is also discussed in the book [SC]. In brief, the prosecutor invited the jury to multiply together the probabilities of six events to show that this number was extremely small and that the defendant, exhibiting these events, was therefore likely guilty. The events, however, are likely not independent. Do you also see how this case illustrates the Prosecutor's Fallacy, discussed earlier?

Exercises

3.18. Alice and Bob are playing darts. The probability that Alice hits the bullseye is .1. The probability that they both hit the bullseye is .02. Assuming that Alice's shot has no effect on Bob's, what is the probability that Bob hits the bullseye?

3.19. A fair die is rolled. Let A be the event "the number shown is greater than 3" and B the event "the number shown is odd". Are A and B independent events? Explain mathematically, rather than intuitively.

3.20. If A and B are independent events with $P(A) = \frac{1}{2}$ and $P(B) = \frac{1}{3}$, what is $(A|B)$? Explain.
3.21. In a family with two children, let A be the event "both children are boys", B the event "the oldest child is a boy" and C the event "there is exactly one boy". Are A and B independent events? Explain mathematically, rather than intuitively.
3.22. In the previous problem, are events A and C independent? Again, give a mathematical explanation.
3.23. If A and B are independent events, with $P(A) = .3$ and $P(B) = .4$, determine the probabilities of the events "A and B" and "A or B".
3.24. If A and B are independent events, and A has probability .1 and "A or B" has probability .2, what is the probability of B?
3.25. Read the Wikipedia page on *People v Collins* (or some other Internet account of this case) and prepare a brief essay, summarizing, in your own words, the misuse of independent events in this case.

3.4 The Monty Hall Problem

In this section (which we will not use in the rest of the text), we discuss a famous problem in probability theory and its resolution. The problem is interesting, has an amusing history, and has a very counter-intuitive solution. It is also almost certainly the only problem in mathematics that is named after a television game show host. A number of books have been written about it, one of which [Ros] is particularly good and highly recommended.

First, some history. Our discussion begins with a woman named Marilyn vos Savant, who for many years has had a column in Parade Magazine. Her claim to fame is that she was once listed in the Guinness Book of World Records as having the highest IQ in the world. (This category has since been retired.) In her columns, she would often answer questions from readers. (See exercise 3.9.)

One such question, published in a column in 1990, sparked considerable discussion and controversy. This question seems to be inspired by a television game show called Let's Make a Deal, which has been on the air (in several incarnations) for decades. At the time this question was posed to vos Savant, it was hosted by a man, now deceased, named Monty Hall. The format of the show involved Monty selecting people from the audience and offering them various deals. At the end of the show, Monty would offer contestants who had won some money an opportunity to participate in a final deal, which, as the question indicates, involved three doors, behind one of which was a fabulous prize ("car"); behind the other two doors were gag gifts (the "goats").

The question posed was:

> Suppose you're on a game show, and you're given the choice of three doors: Behind one door is a car; behind the others, goats. You pick a door, say No. 1, and the host, who knows what's behind the doors,

Conditional Probability and Independence

opens another door, say No. 3, which has a goat. He then says to you, "Do you want to pick door No. 2?" Is it to your advantage to switch your choice?

Most people, upon hearing this question, answer that there is no advantage to switch doors: after all, they reason, there are two doors left, one of which contains the car and the other of which does not. Since the probability that the door you picked contains the car is ½, you do not improve your probability of getting the car if you switch.

This is not, however, the answer that vos Savant gave: she said you double your probability of success by switching. Reaction from her readers (including mathematicians) was swift and sometimes nasty. Here are just a few of the actual quotes (with sender's name omitted) from mathematicians who wrote in to complain about her answer:

> "You blew it, and you blew it big.... There is enough mathematical illiteracy in this country, and we don't need the world's highest I.Q. propogating more. Shame!"
>
> "May I suggest that you obtain and refer to a standard textbook on probability before you try and answer a question of this type again?"
>
> "Maybe women look at math problems differently than men."

And yet, vos Savant was correct. There are a number of ways to see why, but we will give an argument that uses ideas about conditional probability. We will make explicit two assumptions about the game setup that vos Savant did not explicitly set out in her first response. Specifically, we assume that Monty always offers another door and always offers a door that contains a goat. In the event that there are two doors that contain goats, we assume Monty will select one to open randomly and with equal probability.

With these assumptions in mind, let us analyze the problem. We first suppose that you follow a policy of not switching. In that case, anything that Monty does is irrelevant; you chose door No. 1 and will stick with that choice until the end, ignoring anything that Monty says or does. Since the probability of you having selected the correct door is ⅓, this remains the probability that you will get a car.

Now, however, let us suppose that you follow a policy of switching. Denote by W the event "you win the car", and by D_1 the event "the car is behind door number 1". Events D_2 and D_3 are defined similarly. Now, we know that

$$P(W) = P(W|D_1)P(D_1) + P(W|D_2)P(D_2) + P(W|D_3)P(D_3).$$

The individual probabilities on the right-hand side of this equation are obvious:

$$P(D_1) = P(D_2) = P(D_3) = \frac{1}{3},$$

and (keeping in mind that you have selected door No. 1), $P(W|D_1) = 0$, $P(W|D_2) = P(W|D_3) = 1$. (This is because, by switching, you ultimately select a door other than No. 1, and will therefore definitely lose if the car was in fact behind that door, and will definitely win if the car was behind the other possible door.) Plugging these values into the equation, we get $P(W) = 2/3$, showing that following a policy of switching does indeed double the expected value of getting the car rather than the goat.

Exercises

3.26. There are a number of Monty Hall simulation sites online, including at https://www.mathwarehouse.com/monty-hall-simulation-online or https://www.rossmanchance.com/applets/2021/montyhall/Monty.html. Pick the site of your choice and play ten rounds, following a "don't switch" policy. Keep track of the number of successes you have in these ten rounds. Then play ten more rounds, now following a policy of always switching. Again, keep track of the outcomes. Did your success rate increase by always switching?

3.27. Suppose we vary the Monty Hall problem by now assuming there are four doors, behind one of which there is a car, with a goat behind each of the other three doors. Suppose also that after you make your choice, the host opens another door and reveals a goat, then offers you a chance to switch your choice. Assume that the host randomly selects one of the losing doors to show you and that you randomly select one of the two remaining doors if you switch. What is your probability of getting a car if you switch? What is your probability of getting a car if you don't switch?

4

Craps

The game of craps is another of those casino games that you have likely seen pictures or videos of. When played in a casino, one person (the "shooter") stands in front of a table while many other people stand around at the sides. The shooter rolls a pair of dice, often first blowing on them for luck or saying something along the lines of "baby needs a new pair of shoes!". Many of the bystanders have also made bets of their own, too. There is a lot of shouting or jeering as the dice are rolled. It is a fun game to watch and play, and an added benefit is that, as casino games go, this is one that offers a relatively low (but still positive, of course!) expected value to the casino. The game is also played outside of a casino; there are lots of photos extant of soldiers playing the game outside their barracks. (When played outside of a casino, the term "street craps" is used to describe the game; the players play against each other rather than a casino.)

4.1 Rules of the Game

We first discuss the rules of the game. For the moment, we will ignore the kind of side bets that the bystanders can make and focus exclusively on the shooter's game. The shooter, as noted, rolls two dice, and it is the sum of the dice, rather than the individual die themselves, that determines whether he or she wins or loses. Because of this, when we say "the shooter rolled a 7" in what follows, we mean that the sum of the two dice is 7.

The rules that determine whether the shooter wins or loses are as follows:

- If the shooter rolls a 2, 3, or 12 on the first roll (sometimes called the "come out roll") of the dice, he or she loses immediately.
- If the shooter rolls a 7 or 11 on the first roll of the dice, he or she wins immediately.
- If the shooter, on the first roll of the dice, rolls anything other than a 2, 3, 7, 11, or 12, the number rolled becomes the "point". The shooter then repeatedly rolls the dice until either the point or 7 is rolled. If a 7 is rolled before the point, the shooter loses; if the point is rolled before a 7, the shooter wins.

DOI: 10.1201/9781032659145-4

The game plays even money. If the shooter bets $1 (again, for the sake of simplicity, we will assume that bets are for a dollar each) and wins the round, either on the first throw or a subsequent one, then he or she wins a dollar from the casino (and gets his original bet back, for a total profit of a dollar). If the shooter loses, his or her original bet is lost.

As an illustration, suppose the shooter initially rolls a (4, 5). (See Section 1.2 for the notation.) Since the sum here is 9, which is not one of the "automatic win or lose" numbers, the shooter rolls again. Assume a (4, 4) is rolled. Since the sum here is 8, which is neither the point nor 7, this roll is irrelevant, and the shooter rolls again. Suppose now the shooter rolls (6, 3). This is 9, the point, so the shooter wins the game. If the shooter had instead rolled a (6, 1), he or she would have lost, and if he or she had rolled a (3, 1), yet another roll would have been required.

In the next section, we will discuss the expected value, to the shooter, of a $1 bet at the craps table. To prepare for this and to help the reader assimilate the rules of the game, we will do some rather simple probability calculations. Some of these have appeared previously in the text but will be reprinted here for the sake of convenience.

Example 4.1.1. What is the probability that the shooter rolls a 7 on the first roll?

Solution. If we refer to the sample space whose 36 elements are written out in Section 1.2, we see that of the 36 possible outcomes, six of them are "favorable": (1, 6), (2, 5), (3, 4), (4, 3), (5, 2), and (6, 1). Dividing 6 by 36 yields the answer: $1/6$.

Example 4.1.2. What is the probability that the shooter wins on the first roll?

Solution. For this to happen, the shooter must roll a 7 or 11. There are, as seen in the previous example, six ways to roll a 7; there are two ways to roll an 11 ((4, 5) or (5, 4)). So, there are eight ways to win out of a total of 36 possible outcomes, and the probability of winning on the first roll is therefore $8/36 = 2/9$.

Example 4.1.3. If the shooter initially rolls a (3, 1), what is the probability that the shooter ultimately wins?

Solution. This is example 3.1.3 in the text, but here is the solution again. First, notice that any roll that does not produce a sum of 4 or 7 is simply irrelevant; it adds nothing to the discussion and can be ignored. Thus, the only kind of roll that is of interest to us is one where the sum is either 4 or 7. It therefore follows that the problem may be rephrased as: what is the probability that the sum of a roll is 4, given that it is

Craps

either 4 or 7? Once rephrased, the problem becomes easy to solve. We have already seen that the sample space for a roll of two dice has 36 elements. If we adjust this sample space by removing entries that reflect a sum other than 4 or 7, we are left with 9 elements: (1, 3), (2, 2), (3, 1), (1, 6), (2, 5), (3, 4), (4, 3), (5, 2), and (6, 1). Of these nine elements, three of them are favorable. Therefore, the desired answer is ⅓.

Exercises

4.1. If the shooter rolls a (3, 3) on the initial roll, what is the probability that he or she ultimately wins?
4.2. If the shooter rolls a (3, 3) on the initial roll, what is the probability that the shooter rolls another (3, 3) before rolling a sum of 7?
4.3. What is the probability the shooter establishes a point of 9 on the first roll?
4.4. If the shooter in the previous problem establishes a point of 9 on the first roll, what is the probability that he or she will ultimately win?
4.5. If the shooter in the previous two problems establishes a point of 9 on the first roll, what is, immediately after that point is established, the expected value of the shooter's $1 bet?

4.2 Analysis of the Shooter's Game

This section is largely devoted to a discussion of what the expected value, to the shooter, is of a craps bet. Let us denote by W the event "the shooter wins", and let L denote the complimentary event "the shooter loses". Because the shooter makes a $1 profit on a win and a $1 loss on a loss, the expected value to the shooter is (+1) P(W) − P(L). Since P(L) = 1 − P(W), it suffices to find P(W). This isn't hard, but it does take a certain amount of work.

Determining the probability that the shooter wins requires consideration of a number of cases because there are lots of ways the shooter can win: on the first round by rolling a 7 or 11, or on subsequent rounds after the shooter has established one of several possible points (4, 5, 6, etc.). This is a classic example of the Law of Total Probability, which was discussed in Section 3.2.

The rules of the game make clear that the probability of winning depends on the result of the first roll: if a 2 is rolled, the probability of winning is 0; if a 7 is rolled, the probability of winning is 1; and if, for example, a 4 is rolled, the probability of winning is, as per example 4.1.3 in the previous section, ⅓. Therefore, determining the general probability of winning (before any throws have been made) is a classic illustration of the Law of Total Probability. Let us denote by W the probability of winning; by A_2 the probability of rolling a 2 initially; by A_3 the probability of rolling a 3 initially, and so on. With this notation, the Law of Total Probability states that

$$\begin{aligned} P(W) = \ & P(W|A_2)P(A_2) \\ + \ & P(W|A_3)P(A_3) \\ + \ & P(W|A_4)P(A_4) \\ + \ & P(W|A_5)P(A_5) \\ + \ & P(W|A_6)P(A_6) \\ + \ & P(W|A_7)P(A_7) \\ + \ & P(W|A_8)P(A_8) \\ + \ & P(W|A_9)P(A_9) \\ + \ & P(W|A_{10})P(A_{10}) \\ + \ & P(W|A_{11})P(A_{11}) \\ + \ & P(W|A_{12})P(A_{12}) \end{aligned}$$

So, to compute P(W), we need only compute each of the 22 probabilities that appear on the right-hand side above, plug each of these values into the equation, and do the required multiplications and additions. As it turns out, the first of these chores, computing the probabilities, is fairly simple; we have, in fact, done examples just like this before.

The same method used to solve example 4.1.3 can be used, with very minimal alteration, to find the probability of winning given any other point. If we denote by W the probability of winning, by A_2 the probability of rolling a 2 initially, by A_3 the probability of rolling a 3 initially, and so on, then the reader can (and should!) easily verify that

$$\begin{aligned} P(W|A_2) &= 0 = P(W|A_{12}) = P(W|A_3) \\ P(W|A_7) &= 1 = P(W|A_{11}) \\ P(W|A_4) &= \tfrac{1}{3} = P(W|A_{10}) \\ P(W|A_5) &= \tfrac{2}{5} = P(W|A_9) \\ P(W|A_6) &= \tfrac{5}{11} = P(W|A_8) \end{aligned}$$

Craps

Computing each of the probabilities $P(A_2)$, $P(A_2)$, etc., is also an exercise we have already done—consult example 4.1.1 for $P(A_7)$. The method used in this example works for the other probabilities as well. Again, the reader should verify that

$$
\begin{aligned}
P(A_2) &= 1/36 = P(A_{12}) \\
P(A_3) &= 2/36 = P(A_{11}) \\
P(A_4) &= 3/36 = P(A_{10}) \\
P(A_5) &= 4/36 = P(A_9) \\
P(A_6) &= 5/36 = P(A_8) \\
P(A_7) &= 6/36
\end{aligned}
$$

It now remains only to plug these values into the 11-line formula for $P(W)$ appearing earlier in this section, do the multiplication and addition, and arrive at the numerical value of $P(W)$. This is, as may be imagined, a tedious and unpleasant chore, but fortunately, it is one that has been done before, and the answer appears in many books. So, rather than reinvent the wheel, we will simply rely on these other sources and quote the answer: $P(W) = 244/495$, which is (approximately) .4929. It follows from the "complimentary event" rule that the probability that the shooter loses is approximately $1 - .4929 = .5071$.

Something interesting is going on here. Note that the probability that the shooter wins is very close to ½, making this a pretty good bet for the shooter. In fact, we can determine how good by looking at the shooter's expected value: since the shooter wins with probability .4929 (outcome: +1) and loses (outcome: −1) with probability .5071, the expected value is simply $.4929 - .5071 = -.0142$, meaning that the shooter loses about 1.4 cents on the dollar in the long run. Contrast this with a roulette bet, where the casino takes more than 5 cents per-dollar—almost four times as much.

Exercises

4.6. Verify that $P(W|A_4) = 1/3 = P(W|A_{10})$.
4.7. Verify that $P(W|A_5) = 2/5 = P(W|A_9)$.
4.8. Verify that $P(W|A_6) = 5/36 = P(W|A_8)$.
4.9. Verify the statements above for $P(A_2), \ldots, P(A_{12})$.

4.3 Other Bets

At any casino craps table, there will be a lot of bystanders standing around while the shooter rolls the dice. Since the casino would much rather have these people betting than just standing there, it provides for a number of bets that they can make. The most obvious ones concern whether the shooter will, or will not, win.

A "pass" bet is simply a bet, at even money, that the shooter wins; a "don't pass" bet is a bet that he or she loses. Since the payoff on a winning pass bet is the same to a bystander as it is to the shooter, and since the probability of success is also the same, it follows that the expected value on a pass bet is exactly the same as it is for the shooter to win: $-.0142$. But what about the expected value on a don't pass bet? Since the probability that the shooter loses (and therefore a don't pass bet wins) is greater than ½, this means that the expected value of a don't pass is positive. Specifically, it is $.5071 - .4929 = .0142$.

Obviously, this (or any other bet with a positive expected value to the gambler) is intolerable for a casino. To provide the casino with a positive expected value, some adjustments to the payout must be made. Many casinos have come up with the following: if the shooter loses by rolling a 12 on the initial roll, then, instead of paying a "don't pass" bettor the amount of his or her bet, the casino simply calls it even and does not pay anything. From the standpoint of the gambler, it's a draw.

How does this affect the expected value of a "don't pass" bet? The probability of rolling a 12 is $\frac{1}{36}$ (which is approximately .0278); to the sum defining the probability of this bet, it now contributes 0. Therefore, the expected value of this bet to the gambler is reduced by roughly .0278. With this new rule in place, therefore, the expected value of the "don't pass" bet is $.0142 - .0278$, or $-.0136$. So, this change in payout now produces a bet that is acceptable to the casino, providing it, in the long run, roughly 1.4 cents on every dollar bet. This expected value is very close to, but slightly smaller than, the expected value to the gambler on a "pass" bet, so if you're going to bet "pass" or "don't pass", the slightly better bet is "don't pass". (Psychological factors, however, may play a role in the choice of bet; some people may not wish to bet against a spouse or other loved one, for example.)

There are other bets a bystander can play; we'll discuss a couple in the text and a few more in the exercises. The bets are specified on a craps layout that gamblers place chips on.

Two of these bets are "Come" and "Don't Come" bets; these are just like "Pass" and "Don't Pass" except that when these bets are placed, the next roll of the dice by the shooter becomes, for purposes of these bets, the new come-out roll. The mathematics of these bets does not differ from the mathematics of the Pass/Don't Pass bets, and therefore we will not discuss these bets further here.

Another bystander bet is the "Big 8", in which the gambler bets that an 8 will appear before a 7. This bet pays even money, so a winner makes a

$1 profit on a $1 bet. The expected value to the gambler is easy to compute. The probability that an 8 comes up before a 7 is $5/11$ (we noted this in the previous section), so the gambler's expected value is $(+1)(5/11) - 6/11 = -1/11$, which is approximately $-.0909$. Therefore, the casino wins, in the long run, around 9 cents per-dollar on this bet. This makes this a terrible bet for the gambler, *much* worse than pass/don't pass and even worse than roulette. (We will see an even worse bet in a second.) There is also a "Big 6" bet, with just about the same rules, payout, and expected value of the "Big 8", the only difference being that now the gambler bets a 6 shows up instead of an 8.

There is also a "Hard 8" bet. Here, the gambler bets that a "hard 8" (i.e., double 4s) shows up before a 7 or any other kind of 8. A winner here makes a nine-dollar profit on a one-dollar bet. Using the method of the last section, we can easily show (do so!) that the probability of winning this bet is $1/11$, so a simple expected value calculation shows the expected value to be $-1/11$, the same as in a Big 8 bet.

Some bystander bets do not involve the question of whether the shooter wins or loses but only involve the very next roll of the dice. For example, a "Big Red" or "Any 7" bet is that the very next roll is a 7. A winning Big Red bet pays a $4 profit on a $1 bet. The expected value to the gambler for this bet is also easy to compute. We know that the probability of rolling a 7 on a single bet is $1/6$ and, therefore, the probability of not doing so is $5/6$. It follows that the expected value to the gambler is $4(1/6) - 5/6 = -1/6$, which is around $-.1667$. In other words, on a Big Red bet, the gambler loses, in the long run, almost 17 cents on every dollar bet. This makes the Big Red a terrible bet, one of the very worst that can be made in a casino.

We close this section with a brief look at the concept of Free Odds, which no discussion of craps bets should ignore. The reader may wish to quickly review Section 1.7 to brush up on the terminology that we are about to use.

We first make a preliminary observation. If a casino offers, as a payout rate, the actual mathematical odds of a bet, then that bet is fair: the expected value is 0 for both the casino and the player. Let us look at an example. If you roll two dice, the probability of getting a sum of 12 is $1/36$. In "odds" terminology, the odds against doing this are 35:1. Suppose the casino offered a 35-dollar profit on every $1 bet that this sum occurs. Then the expected value to the gambler is $35(1/36) + (-1)35/36 = 0$. Other examples can also be constructed to illustrate this point—there is no advantage to anybody in getting paid off in accordance with the actual odds of an event occurring. This observation is at the heart of the concept of a free-odds bet.

A shooter in a casino craps game who has neither won nor lost on the first round and who has therefore established a point may, in most casinos, then take out a second bet in the same amount at a more favorable payout rate—specifically, at the true mathematical odds. This second bet, as we have just observed, offers an expected value of 0 to both parties; when combined with the first bet, it lowers the per-dollar expected value to the casino and raises it for the gambler.

As an example, suppose the shooter has bet $1 and has established a point of 10. At this point, we see, using the technique already developed, that the probability that the shooter will win (i.e., the probability that the shooter will roll a 10 before a 7) is $1/3$. In other words, the mathematical odds against this happening are 2:1. The expected value of this even-money bet at this point is $1/3(+1) + 2/3(-1) = -1/3$, meaning that the casino takes more than 33 cents per-dollar on average in this situation. This is not a great place for the shooter to be in.

However, the shooter can now take out a new free-odds $1 bet at a payout rate of 2:1 (the mathematical odds against winning). On this new bet, therefore, the bettor makes a $2 profit if the point is made and a one-dollar loss otherwise. If we combine the two bets (which will either be both won or both lost), then we see that the probability of winning is still $1/3$, but the payout on the combined $2 bet is $3 if the point is made and a two-dollar loss otherwise. So, the expected value is now $1/3(+3) + 2/3(-2) = -1/3$. This is the same fraction as before, but it is now the expected value on a *two*-dollar bet rather than a one-dollar bet; so, on a one-dollar bet, the casino's house advantage has been cut in half.

Exercises

4.10. In an 11 bet, the casino pays 16:1 if the next roll of the dice results in a sum of 11; otherwise, the bettor loses. Compute the expected value to the bettor of this bet.

4.11. Another bet offered at a craps table is a so-called Field Bet, the details of which vary from casino to casino. In one version, the bettor bets that on the next roll of the dice, he or she will roll a sum of 2, 3, 4, 9, 10, 11, or 12. If the bettor does this, the casino pays even money unless a 2 or 12 is rolled, in which case the casino pays 2:1. If any other sum is roll, the bettor loses. Compute the expected value to the bettor of this bet.

4.12. The "7 Point 7" bet is placed right before the first (or come-out) roll of the dice. If the first roll is a 7, the bettor wins at 2:1 payout. If the first roll is a 2, 3, 11, or 12, the bettor loses. If neither of these events happens, then the bettor wins 3:1 if the second roll is a 7, and otherwise loses. What is the expected value to the gambler on this bet?

5

Counting Large Sets: An Introduction to Combinatorics

Up to this point, the various probability calculations that we have had to make involved relatively small sets; the size of the various events (sets) was sufficiently small that they could be counted by hand. But a lot of games in gambling involve much larger sets that cannot be realistically enumerated. For example, suppose you are dealt a hand of, say, five cards (as happens all the time in many variations of the popular card game poker). The total number of such hands is immense, as you should have no trouble imagining for yourself. Trying to do a simple probability problem involving such a hand (for example, what is the probability that the hand you have been dealt contains all hearts?) would therefore be very difficult unless we can figure out a way of figuring out what the denominator of this fraction is. As it turns out, although we cannot enumerate all the five-card hands, we can, with just a little bit more of mathematical background, figure out quite easily how many such hands exist. This chapter will deal with these questions.

5.1 Two Counting Rules

We first introduce two basic, and intuitively obvious, rules, that are the basis for most of what we do in this chapter.

The first of these rules is called the *addition principle*. Before stating it, we give a trivial illustration of it to make the statement much more plausible. Suppose you are given a stack of five fiction books and another stack of three non-fiction books. You are told that you can select one book, fiction or non-fiction, to keep. In how many ways can you make this selection? Obviously, the answer is eight; that's the number of books you have to choose from. The point here is that we have added the number of choices you have to make (five books for one category, three for the other category). We can formalize this intuition as follows:

Addition Principle: If one task can be performed in M ways and a second task can be formed in N ways, then the number of ways of performing *either* the first *or* the second task is $M + N$.

Although we have stated this principle for two tasks, it generalizes naturally to three or more. Here is an example:

Example 5.1.1. At a restaurant, Jane's mother tells her that she can have either pie, cake, or ice cream for dessert. The restaurant serves three different kinds of pie, two different kinds of cake, and five different kinds of ice cream. How many different desserts are available to Jane?

Solution. There are three ways to do the first task (selecting what flavor of pie), two ways to do the second (selecting the flavor of cake), and five ways to do the third (selecting the flavor of ice cream), so Jane has a total of ten choices.

We now state the second major counting principle that we will rely on. Actually, this has already been stated (in Section 1.2), but we repeat it here for convenience.

Multiplication Principle: If a first task can be done in M possible ways and, regardless of how the first task was done, a second task can be done in N possible ways, then there are $M \times N$ ways to do the two tasks, one after the other.

As with the addition principle, this can be generalized to three or more tasks.

Example 5.1.2. A guest at a catered reception is given a menu with three appetizers, four entrées, and two desserts. She can order one item from each category. In many ways, can she create a meal?

Solution. By a straightforward application of the Multiplication Principle, there are $3 \times 4 \times 2 = 24$ possible meals the guest can construct.

Example 5.1.3. Under the circumstances of the preceding problem, assume a guest can order one entrée and either one appetizer or one dessert. In how many ways can she create a meal?

Solution. By the addition principle, the guest can select an appetizer or dessert in $3 + 2 = 5$ ways. Having made this selection, the guest can select an entrée in four ways. So, the total number of ways of creating a meal is $5 \times 4 = 20$ ways.

Example 5.1.4. Mr. and Mrs. Smith, and their son and daughter, want to pose for a picture with them standing in a line. In many ways, can they arrange themselves, left to right, for the photo?

Solution. Think of this as filling in four slots. The leftmost slot can be filled with any of the four members of the family. The next slot can be filled with any of the three remaining members. Then the third slot can be filled with any of the two remaining members, and the one remaining member must occupy the right-most slot. So, the total number of ways of arranging the Smith family in a row is $4 \times 3 \times 2 \times 1 = 24$.

Counting Large Sets: An Introduction to Combinatorics

Numbers like $4 \times 3 \times 2 \times 1$ come up a lot in mathematics—so much so, in fact, that they have their own notation. In general, if we start with a positive integer N and then multiply all the positive integers from N to 1, the resulting product is called "N factorial" and denoted by "$N!$". The number $4 \times 3 \times 2 \times 1$ is, of course, in this notation, $4!$. Likewise, $6!$ is $6 \times 5 \times 4 \times 3 \times 2 \times 1 = 720$. We also define (for reasons that we will discuss at the end of the next section) $0!$ to be 1. We will not define "$N!$", in this book, for negative integers N.

It is helpful to think of $N!$ as the number of ways of lining up N different objects in a row. For example, suppose an experiment is conducted in which ten people run a foot race. Assuming no ties, the total number of possible outcomes in this race is $10!$. Do you see why? The number $10!$, by the way, is equal to $3{,}628{,}800$, as you can readily verify with a calculator. This shows that factorials grow very large very quickly.

Let us return to the Smith family and their photo.

Example 5.1.5. In how many ways can the Smith family line up for a photo if a parent is to be at each end of the line?

Solution. Again, think of this problem as filling in slots. The slot on the left must be filled in with a parent, so there are two choices here. Having made a choice, there is only one parent left to fill in the slot on the right. The second slot from the left can be filled in with a child; there are again two choices for which one. And the third slot from the left can then only be filled in one way since there is only one child left. So, the total number of ways to fill in the slots (i.e., line up the Smiths) is four.

Example 5.1.6. In the situation of example 5.1.4, how many ways are there to line the Smiths up if the two parents want to be next to each other?

Solution. Here's a clever way to solve this. Think of filling in slots with one of the letters M (mother), F (father), S (son), or D (daughter). If the mother and father want to be next to each other, we must have either MF or FM appearing in the list of letters. Let us first consider the case MF: how many ways can we arrange the symbols S, D, and MF in a row? Clearly, the answer is $3! = 6$. Likewise, there are six ways to line them up if the father is to be the left of the mother (i.e., we have FM instead of MF). By the addition principle, there are 12 ways to line them up with the parents next to each other.

Example 5.1.7. In how many ways can we line up the Smith family if the parents are not adjacent to one another?

Solution. There are 24 total possible ways of lining up the members of the family (example 5.1.4), and in 12 of them, the parents *will* be next to each other. So, there must be $24 - 12 = 12$ ways of lining them up so that they will *not* be adjacent to each other.

Example 5.1.8. In how many ways can we line up the members of the Smith family if the two females (M and D) and the two males (F and S) want to stand next to each other?

Solution. In our row, we will either have the males occupy the first two slots and the females the last two slots, or vice versa. Suppose first that the males occupy the first two slots. We have two choices (F or S) for the first slot, and then only one choice for the second, since there is only one male left. Likewise, for the third slot, we can choose two of the females (M or D), and then we have one choice for the final slot. By the multiplication principle, there are thus four ways to arrange the Smith family in this situation. The same reasoning shows that there are four ways to arrange the Smith family in the case where the two females occupy the first two slots. So, by the addition principle, there are $4 + 4 = 8$ ways to arrange the Smith family so as to have the males and females both together.

We now leave the Smith family and give a simple application of the ideas we have been discussing to basic set theory.

Example 5.1.9. How many subsets does a set A with n elements have?

Solution. For each one of the n elements of A, we ask: do we want this element in the subset? Every subset of A is uniquely determined by the choices we make. We have two choices for each element of A: either "yes" or "no". The total number of ways to make our choices, therefore, which is the total number of subsets, is 2 times itself n times, or 2^n.

Another application of these ideas is the rearrangement of letters to form words. In the discussion that follows, the term "code word" will refer to any string of English letters in a specified order, whether or not that string forms an actual word in English or any other language. We begin with a very simple question: in how many ways can we form a code word by rearranging the letters in the name MARK? The answer here, of course, is $4! = 120$; this is just example 5.1.4., rephrased. But now let's ask a slightly different question: in how many ways can we form a code word by rearranging the letters in the word BOYHOOD? By analogy with the first question, we may be tempted to answer $7! = 5{,}040$. But, in fact, this is incorrect because the letter O in BOYHOOD is repeated. Here's why: suppose we initially think of the three Os in the word as distinct; let us call them O_1, O_2, and O_3 to tell them apart. Then, of the $7!$ ways we can arrange the now-distinct letters of the word, we will have (among many other codewords) the following:

$$O_2O_1O_3BYHD$$
$$O_1O_2O_3BYHD$$
$$O_3O_2O_1BYHD$$

These are all distinct if we think of the Os as being different, but of course, they are not different: they are all the same letter O, and all three of these therefore correspond to OOOBYHD, as do three other words that you are encouraged to find and write down for yourself. In other words, the number $7!$

overcounts the total number of codeword rearrangements by 3! = 6, the total number of ways that we can rearrange the three "subscripted" Os.

What about rearranging the letters of the word TOOT? We start as before with 4! rearrangements of the letters in this world, treating all letters as somehow distinct (or made distinct by the addition of subscripts). We then notice that there are two Os, and rearranging them does not change the codeword, so 4! overcounts by a factor of 2! = 2. But we are not yet done because we also have two repeated Ts. For precisely the same reason as before, we must divide by two to avoid overcounting. So, the total number of distinct codewords here is

$$\frac{4!}{2!2!}$$

which simplifies to six. In general, suppose we have a word with n letters. Suppose one letter is repeated a times, another letter is repeated b times, and a third letter is repeated c times. Then the total number of codewords we can obtain by rearranging the letters of the original word is

$$\frac{n!}{a!b!c!}$$

This generalizes, of course, to the case of four or more repeated letters, but we state the rule for three just for the sake of convenience, and to avoid messy subscripts.

Finally, we look at another famous counter-intuitive result in probability theory called the *birthday problem*. How large must a class of students be to ensure that the probability is greater than half that two students in that class have the same birthday (day and month only, not year)? Most people instinctively give as an answer some number in the range of 150–200. But the surprising answer is 23.

Here's why. (We assume no leap year, that any student is as likely to be born on one date as on any other, and that the birthdays of the students are independent of one another.) To get a handle on the problem, consider first the case where the class has only five students in it. We will compute the probability of the complementary event that all students have different birthdays. The total number of possible assignments of birthdays to students is 365^5 since there are 365 choices of birthdays to every student in the class. The number of ways of assigning a birthday so that no two students have the same one is $365 \times 364 \times 363 \times 362 \times 361$. Dividing the second number by the first gives approximately .973, which means the probability of a repeat birthday is only about .027, or 2.7%. That this number is so small is hardly surprising, given the very small number of students involved.

Motivated by this calculation, we are led to consider, for arbitrary N, the probability of no-repeat birthdays among N people. The denominator of the fraction is 365^N, and the numerator is $365 \times 364 \times \ldots \times (365 - N + 1)$. Using a computer to evaluate this fraction for various values of N, it can be shown that $N = 23$ is the smallest value giving a probability less than ½, which is the same as saying that N is the smallest value for which there is greater than a 50% chance of having repeat birthdays if the number of students is 23 or more.

Of course, this is by no means the same thing as asking, "given a particular date, how many students must there be to ensure a greater than ½ probability that a student in the class will have that day as a birthday?" Unfortunately, that is something that Johnny Carson, the famous former host of the Tonight Show, failed to recognize, as this clip from his show demonstrates: https://www.cornell.edu/video/the-tonight-show-with-johnny-carson-feb-6-1980-excerpt.

Exercises

5.1. Consider the Smith family discussed in this section. Suppose that they are now joined by Mrs. Smith's mother, G. In how many ways can M, F, S, D and G line up for a family photo?

5.2. In how many ways can the new Smith family (see exercise above) line up for a photo if G must be in the middle?

5.3. In how many ways can the new Smith family (see exercise above) line up for a photo if G is in the middle and the parents are at both ends?

5.4. In how many ways can the new Smith family (see exercise above) line up for a photo if the three females (M, D, and G) must all be together?

5.5. Refer to example 5.1.9 above. We have previously noted that the null or empty set $\emptyset$ is a subset of A. What choices produce this subset?

5.6. (Continuation of previous problem.) Let A = {1, 2, 3, 4, 5}. How many subsets of A contain 1 and 2 and do not contain 3?

5.7. (Continuation of previous problems.) How many subsets of A contain either 3 or 4 but not both?

5.8. You have three history books, two math books, and four biology books. In how many ways can you line them up on a shelf if books of the same subject must be together?

5.9. A student takes a ten-question multiple-choice exam. Each question has four possible answers, exactly one of which is correct. If the student answers each question purely randomly, what is the probability that he gets a perfect score on the exam?

5.10. (Continuation of previous question.) What is the probability that the student gets no question correct?

5.11. (Continuation of previous question.) What is the probability that the student gets exactly one question correct?

5.12. A famous story concerns two students who blow off a final exam and then tell the professor that the reason they couldn't take it was because while driving to the exam they got a flat tire. The professor puts the students in separate rooms, says, "here is your final" and then hands them each a piece of paper with a single question on it: "Which tire was flat?" Assuming random responses by each student, what is the probability that their answers match?

5.13. A small child plays with four blocks, numbered 1 through 4. In how many ways can the child stack the blocks so that an even-numbered block is on top?

5.14. (Continuation of previous question.) In how many ways can the child stack the blocks so that even and odd numbered blocks alternate?

5.15. In the popular game Yahtzee, five dice are rolled. A Yahtzee occurs if all dice show the same number. What is the probability that a Yahtzee occurs on a single roll of five dice?

5.16. Without using any electronic assistance, or even paper and pencil, compute $101!/100!$.

5.17. How many code words can be formed by rearranging the letters in the word FLOUT? What about LOOT? What about MISSISSIPPI?

5.2 Permutations and Combinations

The subject matter of this section is best motivated by an example. Suppose the teacher of a ten-person class has to create a three-person committee of students to give a presentation to the principal. How many such committees can the teacher create?

A plausible, but incorrect, approach to this problem might be as follows: the teacher can choose the first person in the committee in 10 ways, the second in 9 ways, and the third in 8 ways, so the total number of ways she can select three people is $10 \times 9 \times 8 = 720$. Why is this wrong, however?

Suppose the teacher first chooses Alice (A), then Betty (B), and then Charlie (C). This is one of the 720 choices she can make. But another of the choices is B first, then C, and then A. Note, however, that the choice ABC and the choice BCA produce the exact same committee—as would the choices CAB, ACB, etc. In other words, the number 720 that was arrived at in the previous paragraph overcounts the number of different committees because multiple choices may lead to the same committee, and it is the number of committees that is relevant here.

Just how much of an overcount is there? In other words, how many different selections of A, B, and C in some order produce the same committee? That's easy enough to answer:

ABC
ACB
BCA
BAC
CAB
CBA

So, the number of selections that lead to the same committee is 6 = 3!. This isn't surprising because given three people (here, A, B, and C), they can be arranged in 3! different ways.

So, the number of different committees is overcounted by a factor of 6 in the calculation that led to the answer 720. Therefore, the actual number of committees is $\frac{720}{6} = 120$. The number on the left can be rewritten as

$$\frac{10 \times 9 \times 8}{3!}$$

or

$$\frac{10!}{7!3!}$$

because $\frac{10!}{7!}$ is, after some fortuitous cancellation, precisely the same number as $10 \times 9 \times 8$. (Do this with a paper and pencil and make sure you see why this is so.)

What we are seeing here is the difference between an *ordered* and *unordered* list. For a committee, order doesn't count. Likewise, in most cases, when a person is dealt cards, order doesn't count either: if you are playing poker and are dealt five cards, you generally don't care which cards came to you in which order; all you care about is what five cards wind up sitting in your hand.

The calculation above can be generalized and leads to the following important principle: the number of k-person committees that can be formed from a total of n people is $\frac{n!}{k!(n-k)!}$. Alternatively, this number expresses the number of k-element subsets of an n-element set (think of forming a committee as selecting the elements to be chosen).

This is another expression that turns up all over the place in mathematics, and it, too, has been given its own special name and notation. It is called "the binomial coefficient of n over k" or "n choose k", and is denoted by $\binom{n}{k}$. Other notations include $C(n, k)$ or $_nC_k$.

The letter C stands for "combination". We will use the term "combination" in this book to refer to an unordered collection of objects; we will use the term "permutation" to refer to an ordered collection.

Let's do some simple calculations. The binomial coefficient $\binom{10}{2}$ is equal to $\frac{10!}{2!8!} = \frac{10 \times 9}{2!} = 45$. For further practice, you should now verify for yourself that the binomial coefficients $\binom{12}{2}$, $\binom{7}{3}$, and $\binom{4}{2}$ are equal to 66, 35, and 6, respectively.

Some examples will illustrate how useful binomial coefficients are to counting problems that, prior to this chapter, may have been thought intractable.

Example 5.2.1. In many varieties of poker, you are dealt five cards from an ordinary 52-card deck. How many possible 5-card hands are there?

Solution. Obviously, this problem cannot be solved by directly counting the number of possible hands. But we know now that the answer is just $\binom{52}{5}$, which, as can be verified electronically, is equal to 2, 598, 960. (This number will come up again later in the book.)

Example 5.2.2. Of all the possible five-card poker hands, how many consist entirely of hearts?

Solution. Here, instead of selecting five cards from the set of all cards (which has 52 elements), we select from the set of hearts (of which there are 13). So, the answer is $\binom{13}{5}$, which, as can be verified electronically, is equal to 1,287. Note: as we will see in the next chapter, a poker hand in which all cards have the same suit is called a flush.

Example 5.2.3. If you are dealt five cards from a 52-card deck, what is the probability that they are all hearts?

Solution. In view of the previous two examples, we only need to divide 1,287 by 2,598,960. A calculator gives the answer: .000495.

Example 5.2.4. What is the total number of five-card hands that contain all four cards of some rank (i.e., four aces, four kings, etc.)?

Solution. Let us count the total number of ways that we can construct such a hand. First, we have to pick the rank that will be represented by four cards in the hand. We can do this in 13 ways. We then put all four cards of this rank in our hands. This leaves one card to put into the hand, and we can do this in any of 48 ways. So, the total number of four-of-a-kind hands is $13 \times 48 = 624$.

The alert reader will notice that this example does not use binomial coefficients. However, it serves as a nice lead-in to the next example, which does.

Example 5.2.5. What is the total number of five-card hands that contain exactly three cards of some rank?

Solution. As in the previous example, we first pick the rank to be repeated. We can do this in 13 ways, as before. Then, having selected the rank, we must select three cards of this rank. Since there are a total of four such cards, we can do this in $\binom{4}{3} = 4$ ways. After adding these cards to the hand, we must pick two more cards from the 48 possible ones that remain. (Why 48 instead of 49, since we have only used three cards? Because we cannot pick a card of the rank that we have already used, since the problem requires the hand to contain exactly three cards of that rank.) We can pick two cards out of 48 in $\binom{48}{2}$ ways. So, the total number of hands having exactly three cards of the same rank is $4 \times 48 \times \binom{48}{2}$. (Note: it is possible that the remaining two cards will have the same rank; the statement of the problem does not exclude this possibility. In poker, though, a hand with three cards of the same rank and two of the same rank is called a full house and is a better hand than a three-of-a-kind, which contains three cards of the same rank and two remaining cards of different ranks. Had the problem asked for the number of three-of-a-kind hands, a different analysis would have been required. We will study poker hands in more detail in the next chapter.)

Example 5.2.6. A computer password consists of four distinct digits (integers from 0 to 9) arranged in increasing order. For example, 0147 is an acceptable computer password, but 4269 is not. How many acceptable computer passwords are there?

Solution. This seems pretty difficult at first, until we realize that any acceptable computer password is uniquely determined by four different digits. After all, given four digits, there is one and only one way to arrange them in increasing order. So, the problem here really calls for the number of ways to select four digits from 10, and the answer to this question is just $\binom{10}{4}$.

When we discussed factorial notation in the previous section, we remarked that we define 0! to be 1, and promised an explanation later on. Now that we know about binomial coefficients, we can make good on that promise. The binomial coefficient $\binom{5}{5}$, for example, is equal to 1, because there is only one way to choose 5 objects from 5. On the other hand, it is, by definition, equal to $\frac{5!}{5!0!} = \frac{1}{0!}$. This requires 0! to be equal to 1.

Exercises

5.18. From a bag containing ten white balls and two black balls, two balls are selected at random. What is the probability that both balls selected are white?

5.19. (Continuation of previous problem.) What is the probability that both balls selected are black?

5.20. (Continuation of previous problem.) What is the probability that both a white ball and a black ball are selected?

5.21. Evaluate, without electronic assistance, each of $\binom{15}{2}$ and $\binom{8}{4}$.

5.22. Explain how you know, without any computation, that $\binom{125}{50} = \binom{125}{75}$.

5.23. Evaluate $\binom{1001}{1000}$ mentally.

5.24. A mathematics department at a certain university has four full professors, five associate professors, and three assistant professors. In how many ways can the Dean form a six-person committee consisting of two people of each rank?

5.25. (Continuation of previous problem.) Now suppose the Dean has to select, from the 12 professors in the department, a five-person committee and a two-person subcommittee of that committee. Count the number of ways she can do this in two ways: from the "top down" and from the "bottom up". You should get two different expressions in binomial coefficients. Now evaluate both expressions and see if they give the same answer.

5.26. A computer password consists of five symbols in a row. Three of these symbols must be upper-case English letters, and two must be digits from 0 to 9, inclusive. Assuming repetition is allowed, how many computer passwords are there? (Begin by selecting the two slots that will be occupied by integers.)

Counting Large Sets: An Introduction to Combinatorics

5.27. Redo the preceding problem, this time under the assumption that repetition is not allowed.

5.28. If you are dealt three cards from an ordinary 52-card deck, what is the probability that your hand will contain one Jack, one Queen, and one King?

5.29. If you are dealt three cards from an ordinary 52-card deck, what is the probability that your hand will contain no aces at all?

5.30. If you are dealt a five-card poker hand, what is the probability that your hand will contain the ace of spades (and possibly, but not necessarily, other aces as well)?

5.31. If you are dealt a five-card poker hand, what is the probability that your hand will contain the ace of spades and no other aces?

6

Poker

Now that we know something about counting large sets; we can look at the game of poker, perhaps one of the most famous of all card games. There are many variations of the game (some well-known, some obscure), but all generally involve the same basic idea of trying to obtain certain favorable sets of cards, known as poker hands. Examples of poker hands include royal flush, four of a kind, full house, etc. We will begin this chapter by defining all the poker hands of interest to us, and then use the techniques of the previous chapter to establish the probability of being dealt each of these hands if you are dealt five cards from a fair deck. We will then look at some variations of the game.

6.1 Poker Hands and Their Probabilities

As noted above, poker players strive to obtain certain "good" configurations of cards, called poker hands. Some of these hands are rare and therefore valuable; others are less rare and therefore less valuable. In this section of the book, we will define the various poker hands that players may hope to get and, at the same time, establish the probability of being dealt one of these hands if you are dealt five cards. It is this probability that explains the "rareness" of the hands—a royal flush, for example, has probability close to 0 of being dealt to a player, and for this reason, a royal flush is the most valuable of all the poker hands.

In what follows, we discuss the various poker hands in descending order of value, starting with a royal flush. Recall that the probability of being dealt any particular hand is a fraction, the denominator of which is the total number of ways of being dealt five cards, and the numerator of which is the total number of ways of being dealt that hand. The denominator in all these examples, therefore, doesn't change: it is $\binom{52}{5}$, or 2,598,960. (See example 5.2.1 in the previous chapter.)

Royal Flush. A royal flush consists of five specific cards (Ace, King, Queen, Jack, and 10), all of the same suit. There are four ways to select the suit that will be common to all the cards, and once we have made that selection, there is only one way to pick the cards: they must be the Ace, King, Queen, Jack, and 10 of that suit. So, there are four royal flushes out of a total of 2,598,960 possible five-card hands. The probability of a royal flush is therefore .0000015, a very small number indeed.

Straight Flush. A straight flush consists of any five consecutively ranked cards, all of the same suit. When we use the term "straight flush" we are implicitly indicating that it is not a royal flush; no sensible person playing poker would say, "I have a straight flush", when in fact he or she has a royal flush. Note also that an Ace can be considered to be either a 1 or something higher than a King. As for the probability of being dealt a straight flush, there are again four ways to select the common suit of the five cards, but this time, having selected the suit, there are ten ways to select the low card in the straight: anything from an Ace to a 10. So, there are 40 straight flushes, but four of these are royal flushes, so the number of straight-but-not-royal flushes is 36 and the probability of being dealt one is obtained by dividing 36 by 2,598,960. This is approximately .000014.

Four of a Kind. The name gives away the definition: this hand consists of four cards of the same rank (four 3s, four Kings, etc.) and a fifth card. For example, 4♥ 4♠ 4♣ 4♦ A♠ is a four of a kind. To compute the probability of being dealt this hand, we first count the total number of such hands. There are 13 ways to select the rank of the card that will appear four times (a 2, 3, etc.), and having made that choice, there are 48 ways to select the fifth card in our hand, which can be anything but the four cards already chosen. So, there are $13 \times 48 = 624$ four of a kind hands; dividing this number by 2,598,960 gives (approximately) .00024, the probability of being dealt a four of a kind.

Full House. A full house consists of three cards of one rank and two cards of another rank. For example, 2♥ 2♠ 2♣ A♦ A♠ is a full house. To count the number of full houses, we can choose the rank of the card that will appear three times in 13 ways (any of the 13 possible ranks in a deck) and, having made that choice, can select three cards of that rank in $\binom{4}{3} = 4$ ways. (Here is an alternate way to think about this: we choose three cards of a given rank by deciding which card of that rank will *not* be in the hand, and there are four ways to make that choice.) Next, we can choose the rank of the card that will appear twice in 12 ways (any of the ranks other than the one we just selected) and, having made that choice, can select two cards of that rank in $\binom{4}{2} = 6$ ways. So, the total number of ways to make our various selections (i.e., the total number of ways of performing these four tasks in order) is the product of these four numbers, or 3,744. Dividing this by 2,598,960 gives the probability of being dealt a full house. This number is approximately .0014.

Flush. A flush consists of five cards of the same suit. For example, 4♥ 7♥ 8♥ J♥ A♥. (As noted earlier, we do not count a royal flush or straight flush as a flush; we are counting here the "ordinary flushes".) There are four ways to select that suit. Having made that choice, we then must select five cards from the 13 cards of that suit, a task that can be accomplished in $\binom{13}{5} = 1,287$ ways. So, there are $4 \times 1,287 = 5,148$ flushes, but 40 of them are straight (or royal) flushes, so there are 5,108 flushes that aren't anything better. Dividing by 2,598,960 gives (approximately) .002, which is the probability of being dealt an ordinary flush.

Straight. A straight consists of five cards of consecutive ranks, with the Ace functioning either as a 1 or as a number above the King. As before, there are ten ways to select the low card for our straight, and once we make this selection, the ranks of the remaining cards are determined. If, for example, we select 5 as the rank of the low card, we know the other four cards in our hand must have ranks 6, 7, 8, and 9. For each of these five ranks, we can select a card of that rank in four ways (one for each suit). So, there are $10 \times 4^5 = 10{,}240$ ways to make all the necessary selections. However, as before, 40 of these straights are straight (or royal) flushes, so we have a total of 10,200 straights that are not anything better. To obtain the probability of being dealt a straight, we divide 10,200 (as usual) by 2,598,960, getting a number that is approximately .0039.

Three of a Kind. A three of a kind consists of three cards of one rank, with the two remaining cards having different ranks. (If the two remaining cards had the same rank, we would have a full house.) An example of a three of a kind is 2♥ 2♠ 2♣ 5♦ A♠. Calculation of the number of these hands takes a bit of thought. There are 13 ways to pick the rank of the card that will be repeated three times, and, having made that selection, we can pick three cards of that rank in four ways. (See the previous discussion of a full house, where we also selected three cards of a given rank.) Now we must select the other two cards, and, as previously noted, they cannot have the same rank. So, we must select two possible ranks to appear, and we can do this in $\binom{12}{2} = 66$ ways. Having made this choice (say, 5 and Q), we can select each of these cards in four ways (there are four fives and four Queens). So, the total number of ways of obtaining this hand is $13 \times 4 \times 66 \times 4 \times 4 = 54{,}912$. The probability of being dealt a three of a kind is therefore 54,912 divided by 2,598,960, which is approximately .0211. So, there is a little more than a one-in-fifty chance of being dealt a three of a kind.

Two Pair. If a five-card hand contains two cards of one rank, two cards of a different rank, and a fifth card of a third rank, that hand is called a two pair. For example, 2♥ 2♠ 5♣ 5♦ A♠. Computing the number of two pair hands also takes a little thought. We must first pick two ranks (to be the ranks of the repeated cards; in the example just given, we would have picked a 2 and a 5). There are $\binom{13}{2} = 78$ ways to do this. Having picked them (say, 2 and 5), we can then select the two 5 cards in $\binom{4}{2} = 6$ ways and, likewise, the two 2 cards in 6 ways. When we select the fifth card, we know that its rank cannot be the same as the two we have already chosen (otherwise we'd have a full house, not two pair) so there are 11 choices here. Having chosen the rank, we can select the suit in four ways. So, the total number of ways we can have two pair is $78 \times 36 \times 11 \times 4 = 123{,}552$, and the probability of being dealt one is roughly .0475. (So, getting two pair is more than 31,000 times more likely than getting a royal flush.) It should be noted that there is a common error made here: some people think the rank of the first pair can be chosen in 13 ways and the rank of the second in 12, so the number of ways of selecting the ranks is 13×12. This overcounts by a factor of two, since selecting, say, a 5 first and then a 2 is no different, as far as our hand goes, than selecting the 2 first and then a 5.

One Pair. A one-pair hand occurs when there are two cards of the same rank (say, two 2s) and three cards of different ranks—different not only from each other but from the rank of the pair. For example, 2♥ 2♠ 5♣ 8♦ A♠ is a one-pair. Let's count the number of such hands. The card that will be repeated will have any of 13 possible ranks; once the rank is chosen, we can choose two cards of that rank in $\binom{4}{2}$ = 6 ways. We now must select 3 ranks from 12 to use for the remaining cards; we can do this in $\binom{12}{3}$ = 220 ways. Once we have selected these three ranks, there are four choices of cards for each one, for a total of $4 \times 4 \times 4 = 64$ choices. The total number of choices, therefore, is $13 \times 6 \times 220 \times 64 = 1{,}098{,}240$; to obtain the probability of being dealt one pair, just divide this number, as usual, by 2,598,960, getting (approximately) .0422.

Nothing. How many five-card poker hands do not even rise to the level of one pair? One way to obtain that number is to (using the previous calculations) add up the number of hands that are one pair or better, and subtract that from 2,598,960, the total number of hands. If you do the boring arithmetic (or save some time by reading a book where somebody else has done it), you get 1,302,540, making "nothing" the most likely hand to be dealt.

We end this section with a brief, just-for-fun discussion of a card game called Maverick Solitaire, first introduced on an old (1957–1962) TV show called *Maverick*. This show (a legend in TV history) chronicled the adventures of two gambler brothers, Bret and Bart Maverick, as they roamed (mostly separately, but occasionally together) the old West, getting into various (often lighthearted) adventures along the way. In an episode entitled "Rope of Cards", Bret bets another person named Pike that if 25 cards are dealt from a fair deck, they can be rearranged into five non-overlapping hands of five cards each, each of them a "pat hand" (according to a quote on the Internet Movie Database, pat hands are described as "straights, flushes, full houses", but one would think that a four of a kind should be considered a pat hand as well). The cards are dealt, and within minutes, Bret achieves this objective. Afterward, asked by Pike how often this can be done, Bret replies "almost every time". In fact, it has been suggested that this can be done about 98% of the time, and empirical studies bear this out. Some sources say that after this episode aired, businesses reported a surge in the sale of decks of cards, presumably so that people could try the game for themselves.

Exercises

6.1. How many full houses have three aces?
6.2. How many "four of a kind" hands have four Queens and an ace?
6.3. What is the probability of being dealt five cards of the same color?
6.4. What is the probability of being dealt four cards of one suit and one card of a different suit?

6.5. What is the probability of being dealt a two-pair hand, where one pair consists of aces?

6.6. In some poker games (such as video poker), a pair only has value if it is "jacks or better", meaning a pair of jacks, queens, kings, or aces. What is the probability of being dealt a "jacks or better" one-pair hand if you are dealt five cards?

6.7. Poker can be played with three-card hands instead of five-card hands. Some of the standard five-card hands, like four of a kind, obviously have no analog in this game, but other hands, like flushes and a pair, clearly do. How many three-card hands are there?

6.8. (Continuation of previous problem.) If you are dealt three cards, what is the probability that you are dealt any kind of flush (including a straight or royal flush)?

6.9. (Continuation of previous problems.) If you are dealt three cards, what is the probability that you are dealt a pair?

6.10. (Continuation of previous problems.) If you are dealt three cards, what is the probability that you are dealt a three of a kind?

6.11. (Continuation of previous problems.) If you are dealt three cards, what is the probability that you are dealt any kind of straight?

6.12. (Continuation of previous problem.) If you are dealt three cards, what is the probability that you are dealt three aces?

6.13. (Continuation of previous problem.) If you are dealt three cards, what is the probability that you are dealt a King, a Queen, and a Jack?

6.14. In a variation of poker in which you are dealt four cards, what is the probability that you will be dealt four of a kind?

6.15. In a variation of poker in which you are dealt four cards, what is the probability that you will be dealt two pair?

6.16. Find a deck of cards, deal out 25 of them, and play a round of Maverick Solitaire. Did you win?

6.2 Video Poker

As interesting as it may be to compute the probability of being dealt a particular poker hand, the hand that you are dealt is not typically dispositive of the game. Most poker games allow for additional activity after the hand has been dealt, specifically discarding some cards and drawing others. There is also an element of bluffing that is often involved, where a person with a bad hand may still win if he or she can convince the other players to fold (give up) under the mistaken belief that they are facing a person with a good hand. The subject of *game theory* (see Chapter 10) can be used to address, to some extent, some mathematical issues involving bluffing, but for the moment we will focus on a poker game where bluffing is simply irrelevant, namely *video poker*, where the gambler's opponent is not another person but a machine.

Poker

The idea behind video poker is simple enough: you put some money into the machine, and it deals you five cards. (You can play online for free, in which case, instead of money, you simply bet points.) You then discard any or all of these cards, which the machine randomly replaces with any of the remaining cards in the deck. Obviously, you are attempting to make the best hand that you can. A table of payouts appears on the machine. Often, you need to get "Jacks or better" to have any payout at all. (This means at least a pair of Jacks.) Here is a table of payouts, taken from a video poker website (247videopoker.org):

Bet	1
Royal Flush	250
Straight Flush	50
4 of a Kind	25
Full House	9
Flush	6
Straight	4
3 of a Kind	3
2 Pair	2
Jacks or Better	1

Keep in mind, of course, that you have already spent 1 unit to even play the machine, so the actual profit on, say, a straight flush is 249 rather than 250. Getting a pair of Jacks or better simply means you break even.

In the rest of the chapter, we will look at some simple calculations that can arise when playing video poker.

Example 6.2.1. You are dealt 2♥ 5♥ 8♥ J♥ A♠. You don't have any real chance of improving this to a straight, but note that you already have four hearts—i.e., you are missing a flush by only one card. So, one realistic thing that you can do is discard the ace, hoping the machine will replace it with some heart. How likely is this to happen?

Solution. This should be easy. There are 47 cards left in the deck. Of these, nine are hearts. So, the probability of being dealt a heart is $9/47$.

Example 6.2.2. You are dealt 2♥ 2♠ 8♥ 8♠ A♦. If you discard the ace, what is the probability that you will make a full house?

Solution. This is similar to the previous problem. The computer will draw one additional card, which it can do in 47 ways, and you will have a full house if and only if that card is a 2 or an 8. Since there are 2 of each of these ranks left in the deck, there are four ways this can happen. The probability of getting a full house is, therefore, $4/47$.

Example 6.2.3. You are dealt 9♥ 9♠ 9♣ 8♠ A♦. If you discard the 8♠ and A♦, what is the probability that you will make four of a kind?

Solution. This example is a little different than the previous two in that here, the computer must select two cards, not just one. Since order doesn't count, this can be done in $\binom{47}{2}$ ways. To get a four of a kind, one of the cards selected must be a 9 (there's only one way to draw a 9, since three out of four of them are already in your hand), and, having done this, there are 46 cards left, so there are a total of 46 ways to make a four of a kind. The probability of doing this, therefore, is 46 divided by $\binom{47}{2}$. A few minutes with pen and paper, or a few seconds with a calculator, reveal that this is roughly .04255. In other words, there is less than a 1 in 20 chance of filling out a four of a kind.

Example 6.2.4. Under the circumstances of the previous example, what is the probability that you will make a full house?

Solution. This one takes a little thought. The only way to create a full house is to draw a pair. You can't draw a pair of 9s because there is only one 9 left in the deck. You can, however, draw a pair of any other rank, but the number of ways to do so depends on whether you are drawing a pair of 8s or aces, or any of the other ten ranks (this is because there are only three 8s and three aces left in the deck, but four of the other ranks).

So, let us proceed. There are, of course (just as before), a total of $\binom{47}{2}$ ways to draw two new cards, so, as in the previous example, this will be our denominator. For our numerator, we ask: in how many ways can we draw a pair of 8s? There are three 8s left in the deck, so there are $\binom{3}{2} = 3$ ways to pick 2 of them. For aces, the result is the same. For a rank that is neither an 8 nor an ace (and of course not a 9 either), there are four "good" cards left in the deck, so we can draw two of them in $\binom{4}{2} = 6$ ways. So, the total number of ways of drawing a pair is $3 + 3 + (10 \times 6) = 66$ ways. Dividing 66 by $\binom{47}{2}$ (which is 1,081) gives .061, the probability of making a full house. This is a more likely event, therefore, than getting four of a kind, but that's not surprising since four of a kind is a more desirable hand.

Example 6.2.5. You are dealt 2♥ 5♥ 8♥ A♣ A♠. Discuss possible strategies.

Solution. Two obvious potential strategies present themselves. One is to keep the two aces, in which case you are at least assured that you won't lose any money, or you can discard them and keep the hearts, hoping for a flush. Let us compute the probability of this latter strategy. The total number of ways for the machine to select two cards is $\binom{47}{2} = 1,081$. The total number of ways to draw two hearts is $\binom{10}{2} = 45$. The probability of getting a flush is, therefore, roughly .041. If we compute the expected value of a flush (rounding .041 to $1/25$) we get (using the payout table above) $6/25$, much less than the guaranteed expected value of 1 that we get by keeping the aces and discarding the hearts. Of course, this analysis is not airtight. If we discard the aces and draw two cards, we might, by sheer dumb luck, get three of a kind or a full house. We were computing above the expected *value of a flush*. The other possibilities do raise the overall expected value of drawing two cards, but discarding the three hearts also gives us a possibility of improving our hands—perhaps, even if we were incredibly lucky, to four of a kind. Conventional wisdom would be to play it safe and keep the aces.

Example 6.2.6. You are dealt A♥ K♥ Q♥ J♥ 7♣. If you discard the 7, what is the probability of winding up with a royal flush? A straight? Any kind of flush?

Solution. You can draw a card in 47 ways. Of these, only one card (10♥) will give you a royal flush, three cards (any non-heart 10) will get you a straight, and nine cards (any heart) will get you a flush. So, the probabilities of a royal flush, straight, and flush are, respectively, $1/47$, $3/47$, and $9/47$.

Exercises

6.17. In example 6.2.6 above, what is the probability that you get a pair that is jacks or better?

6.18. In example 6.2.5 above, if you keep the aces and discard the remaining three cards, what is the probability that you will end up with four of a kind?

6.19. If you are dealt A♥ K♥ Q♥ 4♥ 7♣ and discard the 4 and 7, what is the probability that you will wind up with a royal flush?

6.20. If, in the previous problem, you again discard the 4 and the 7, what is the probability that you will make a three of a kind?

6.21. You are dealt K♥ K♣ 3♣ 6♥ J. If you keep the two Kings and discard and replace everything else, what is the best possible hand that you can wind up with, and what is the probability that you succeed in doing so?

6.22. You are dealt 7♥ 8♦ 9♥ J♣ Q♦. If you discard the Queen, what is the probability of getting a pair of Jacks or better?

6.23. In the previous problem, if you again discard the Queen, what is the probability of getting a straight?

6.24. In the previous two problems, why is discarding the Queen not a sensible thing to do?

6.25. If you are dealt 4♥ 5♠ 6♣ 7♥ A♠, what is the probability of making a straight if you discard the ace?

6.26. If you are dealt 4♥ 5♠ 7♣ 8♥ A♠, what is the probability of making a straight if you discard the ace? Why is this problem mathematically different from the previous one? (You may have heard of the phrase "Never draw to an inside straight.")

6.3 Texas Hold 'Em

Texas Hold 'Em is one of the most popular, if not *the* most popular, current poker game. In most poker tournaments, the game that will be played is almost always Texas Hold 'Em. Unlike many other varieties of poker, Hold 'Em is based on the idea of sharing cards. The game works as follows: all the players at the table are dealt two cards, face down. Based on this beginning, a round of betting ensues, with each person calling, raising, or folding. After the round of betting has ended, three more cards (the "flop") are dealt face-up; these are "common cards" and can be used by any of the players to build up a hand. Another round of betting then ensues, followed by the deal of a fourth common card, this one called the "turn". Then, after yet another round of betting, a fifth common card, called the "river", is dealt. There is then still another round of betting, after which those players who have not yet dropped out reveal the cards, with, of course, the player with the highest poker hand winning.

As an example, suppose player 1 is dealt an A♥ 4♠ and player 2 is dealt K♠ Q♣. The flop is 5♠ A♠ J♥, the turn is A♣, and the river is Q♠. Player 1's three aces beat Player 2's two pair.

Here are some simple probability examples involving Texas Hold 'Em.

Example 6.3.1. At the very beginning of the game, before any cards have been dealt, what is the probability that you will be dealt two aces? (This is the best possible opening hand; it is called "pocket aces".)

Solution. There are 52 cards, so there are a total of $\binom{52}{2} = 1{,}326$ two-card hands that can be dealt. There are a total of $\binom{4}{2} = 6$ ways to select two aces from the four available, so the probability of being dealt two aces is $6/1{,}326$, which is roughly .0045.

Example 6.3.2. You are watching a game of Hold 'Em on TV. The players are Alan and Bob. You and the other TV viewers can see all the cards that have been dealt, and you see that Alan is initially dealt A♥ A♠ and Bob has been dealt 2♠ 3♠. The flop is A♣ A♦ 5♠. What is the probability that Bob ultimately wins?

Solution. Alan, of course, already has an excellent hand: four aces. The only way Bob can beat this and win is if he gets a straight flush on the turn and river. The only way this can happen is if the last two cards are a four and six of spades. Now, seven cards have been revealed, so 45 cards remain; there are thus $\binom{45}{2} = 990$ ways to choose two cards from these remaining ones. There is only one way to choose the two specific cards Bob needs. So, the probability that he will ultimately beat Alan's four of a kind is $1/990$.

Example 6.3.3. You are dealt 2♠ 3♠. The flop is A♣ A♦ 9♠. What is the probability that, after the turn and river are dealt, you will have a flush?

Solution. Two cards remain to be dealt, and they must both be spades for you to get a flush. There are 47 cards left in the deck, so the total number of ways to select two of them is $\binom{47}{2} = 1{,}081$. There are ten spades remaining in the deck, so the total number of ways choosing two of them is $\binom{10}{2} = 45$. So, the probability that, of the two cards that you will be dealt, both will be spades is $45/1{,}081$.

For readers who wish to read more about probability in the context of Texas Hold 'Em, the book *Introduction to Probability with Texas Hold 'Em Examples* by Schoenberg [Sch] is highly recommended.

Exercises

6.27. What is the probability that the cards dealt to you and your sole opponent, as well as the flop, turn, and river, are all clubs?

6.28. You are initially dealt A♠ A♥. What is the probability that after the flop has been dealt, you will have four of a kind?

6.29. In the previous question, what is the probability that you will have four aces after the flop, river, and turn have been dealt?

6.30. You are dealt A♠ A♥. The flop is 2♥ 3♠ 5♠. Write down five different two-card hands that, if your opponent had been dealt them at the start of the game, would mean that he was now, at this point in the game, beating you? (Varying the suits does not make a hand "different".)

6.31. In example 6.3.3 above, suppose the turn is dealt and is J♠. Now what is the probability that you, at the end of the dealing, have a flush?

6.32. In example 6.3.3 above, suppose the turn is dealt and is A♥. Now what is the probability that you, at the end of the dealing, have four of a kind?

6.33. In exercise 6.2.7 above, what is the probability that at the end of the dealing, you will have a full house?

6.34. You are dealt 6♦2♦. The five common cards are 2♣ 4♣ 6♥ 6♣ 6♠. What do you have? Can your opponent be holding any hand that beats you?

7
Lotteries and Keno

It is likely that most readers of this book have some experience with the concept of a lottery, which, for our purposes, will refer to a game where a player, in exchange for the payment of a small amount of money, is allowed to pick a number or numbers, in the hope that they will match numbers selected later. A church raffle, for example, can be thought of a lottery (you buy a numbered raffle ticket and hope that it matches one selected later), as is, of course, the lottery ticket that many people buy every day at a convenience store or supermarket. Another game with strong lottery overtones is called Keno and is played at many casinos, though its popularity may be waning. We will not penetrate very deeply into these topics, contenting ourselves with an overview of some of the basic mathematics behind them, but lotteries and Keno offer enough interesting mathematics to fill a book—as, for example, [Bol1].

7.1 Lotteries and Powerball

We first study lotteries, particularly government-run ones. Historically, these date back millennia. Even in the United States, they were present in colonial America and predate this country's independence from England. After independence, states used lotteries to raise money for public purposes; for example, southern states used lotteries to help provide money for the Civil War.

By the time the 20th century dawned, lotteries had lost much of their allure and were illegal in most states. This opened the door for *illegal* lotteries, specifically the infamous "numbers racket" in the early 1900s, which provided a huge windfall for organized crime. It worked as follows: the bettor would select three digits in some order. A winning number was then announced, typically by referring to some random and unimpeachable source, such as the last three digits of the U.S. Treasury balance, taken from a newspaper. A bettor who matched (counting order) all three numbers won a 600-to-1 payoff. The bets themselves could be very small; in some cases, they could be as low as a penny. This, unfortunately, made the game very attractive to people who could least afford it. During the Depression, for example, this game became a huge money-maker for organized crime.

Having mentioned this game, let us see just how big a money-maker it is. We will compute the expected value of this game on a $1 bet. The possible

Lotteries and Keno

outcomes of this bet are −1 or +599 (you win 600 dollars but have already spent a dollar). The probability of winning is $1/1000$; this is because there are ten ways to select each digit, and hence 1,000 ways to guess three of them in order; only one of these guesses produces the lucky number. The probability of losing, therefore, is $999/1000$. Using these figures, it is easy to compute that the expected value of this bet is −.4; to put it another way, the "house" (i.e., crime) takes 40 cents of every dollar bet. None of the gambling games studied so far in this book are anywhere near as disadvantageous to the bettor.

Back in 1959, a legendary TV show called *The Untouchables* first aired, starring Robert Stack as Eliot Ness, a federal agent in the 1930s. One episode was entitled "You Can't Pick the Number" and involved Ness' attempt to curtail the numbers racket. In one amusing scene (we'll see why it's amusing very shortly), somebody asks Ness what the harm is in a lottery. He replies, "You give them one chance in a thousand to guess a number. That's not a lottery, it's a swindle." We will revisit this comment shortly.

Over time, state officials began to realize that lotteries offered an easy way to raise revenue without resorting to the politically toxic tactic of raising taxes. In Iowa (where the author lives), then-Governor Branstad, after vetoing lottery legislation several times, finally bent to public opinion and, in April 1985, signed legislation providing for a state-run lottery. Lottery games began in Iowa in August 1985 (with a kickoff at the Iowa State Fair), and have, over the years, grown in scope. The webpage for the Iowa Lottery lists several games that can be played, one of which is called Pick 3. In this game, the bettor selects three digits with the goal of matching them against a number chosen by the lottery at random. The bettor also gets to select the method of play, including straight (match them in the exact order); box (match all three numbers in either order); and front or back pair (match the first (or last) two numbers in exact order). Of course, the "exact" game is precisely the same game as referenced in *The Untouchables* TV show mentioned above and has the same payout. So, the very game that was referred to as a "swindle" on that show is now a state-sanctioned lottery game. Of course, the bettor in the state game does at least have the satisfaction of knowing that the state's exorbitant proceeds are used for public projects rather than more nefarious ones.

We now turn to one of the most famous lottery games, one that Iowa plays in conjunction with (as of this writing) 44 other states, the District of Columbia, Puerto Rico, and the US Virgin Islands. The fact that there is one single lottery for which tickets are purchased all over the country makes the jackpot very large, and in fact, in 2022, a bettor in California won a 2.1 billion dollar jackpot.

The basic idea of the game is fairly simple. A Powerball ticket costs $2, and entitles the bettor to select, first, five different (unordered) numbers, each between 1 and 69, and then a separate Powerball number that is between 1 and 26. The Powerball authorities select the winning number by using balls with numbers on them: one machine bounces 69 white balls around and eventually selects five of them; another machine bounces 26 red balls around and winds up selecting one of them.

There are variations on this basic idea (such as Power Play and Double Play), but we will not discuss them here. We will content ourselves with a quick look at some of the various probabilities connected with a basic Powerball drawing as described above.

After the numbers have been drawn, if anybody has matched all the white ball numbers (order doesn't count) and the red Powerball number, that person wins the jackpot prize, which keeps growing as long as there has been no winner. If there are multiple winners, the prize is split between them.

Example 7.1.1. What is the probability of winning the Powerball jackpot?

Solution. As always, this is the total number of ways of obtaining a favorable outcome (i.e., the total number of ways of filling out a ticket to match the selected numbers) divided by the total number of ways of filling out a ticket. We will compute these numbers under the assumption that order doesn't count. The first of these numbers, the numerator of our fraction, is 1: there is only one way to select the five "correct" numbers and then select the designated Powerball number. As for the denominator, the number of ways to select five numbers from 69 is $\binom{69}{5}$, and the total number of ways of selecting one Powerball number is 26. So, the total number of ways of filling out a Powerball card is, by the product rule, $\binom{69}{5} \times 26$, which you can, with the aid of a calculator or computer, determine to be 292,201,338. Hence, the probability of winning the Powerball jackpot is $1/292,201,338$. To give you an idea of just how small a number this is, consider that the probability of being hit by a meteor has been estimated, in the book *Chance Rules* by Everitt, to be about $1/250,000,000$. In other words, you have a greater likelihood of being hit by a meteor than you do of winning the Powerball jackpot.

What if we did the previous calculation by counting the number of ordered selections? The denominator would now be $69 \times 68 \times 67 \times 66 \times 65 \times 26$. The numerator, however, would now be 5! = 120 because we can select the five winning numbers in 5! ways (keeping in mind that we only have to match the set of winning numbers). It is easy to see, of course, that these answers are the same.

Of course, you don't have to win the jackpot to get some return on your Powerball ticket investment. Smaller prizes are awarded to people who match some, but not all, of their numbers. For example, a person who matches four out of the five white balls and the Powerball earns $50,000.

Example 7.1.2. What is the probability of winning this payout?

Solution. There are $\binom{5}{4} = 5$ ways to select four winning numbers out of the five selected by the lottery. Having selected these four, there are 64 ways to select a number from the "losing" 64. There is only one way to pick the red Powerball. Thus, there are 320 ways to create a ticket that wins $50,000. Dividing by 292,201,338, we see that the probability of winning $50,000 is $320/292201338$, in agreement with the approximate answer given on the Iowa Lottery homepage.

Lotteries and Keno 71

A person who matches only the red Powerball and none of the white balls earns $4. As our last example of this section, we compute the probability of that event occurring.

Example 7.1.3. What is the probability of matching only the Powerball number?

Solution. There are $\binom{64}{5} = 7{,}624{,}512$ ways of selecting five numbers from the 64 "losing" ones, and one way to select the winning the Powerball number. Thus, the probability of matching only the Powerball number is $7{,}624{,}512/292{,}201{,}338$. This agrees with the approximate number given on the Iowa Lottery webpage.

We end this section by saying a few words about the expected value, to the gambler, of a Powerball ticket purchase. If the jackpot prize exceeds 292,201,338, does this mean that the bettor enjoys a positive expected value? The answer is no. When we discussed the numbers racket earlier, it was easy to determine the expected value of the game to the bettor, but that was because the payout for a winning ticket was known. In the case of Powerball, that's not the case: it is impossible to know, prior to the draw, what the exact payout will be. For one thing, if more than one person wins, the jackpot is split among the winners. Also, even in the event that there is only one winner, that winner will not receive the full amount of the jackpot. He may receive an annuity providing for payments over a period of years, or he may instead get a reduced award. And the government will take taxes out of the payout up front as well.

Exercises

7.1. According to the Florida Lottery webpage, their version of the Pick 3 straight game has a payout of 500:1 rather than 600:1. Compute the expected value to the bettor of a $1 bet on this game.

7.2. Suppose that the Powerball rules were changed so as to require that the player match the five white ball numbers in a specific order in order to win the jackpot. What would the probability of winning the jackpot be then?

7.3. If a Powerball player matches the red Powerball number and one out of the five white numbers, he or she wins $4 (the same as if he or she had only matched the Powerball). What is the probability of this occurring?

7.4. What is the probability of matching three out of the five white numbers but not the Powerball?

7.5. Another multi-state lottery game is Mega Millions, in which the player picks five (unordered) white-ball numbers from 1 to 70, and one gold-ball (the Mega ball) number from 1 to 25. If you match all numbers, you win the Jackpot. Which has greater probability, winning the Jackpot in Powerball or in Mega Millions?

7.6. An elementary school puts on a carnival to help raise money for new library books. One of the games at the carnival is a raffle, where 100 people purchase a raffle ticket for $1 each. The school then selects one ticket for a first prize of 30 dollars and, after that, a second ticket for a second prize of 10 dollars. At the time a ticket is purchased, before any drawing has taken place, what is the expected value of a $1 ticket?

7.7. (Continuation of previous exercise.) Now assume that the first prize ticket has been selected and that John, one of the ticket purchasers, did not win. Now, what is the expected value of his ticket?

7.8. Mary buys a Powerball ticket every week and always uses the numbers 1, 2, 3, 4, and 5 as her white ball numbers, and then selects 6 as her red ball number. Her husband John says this is a bad choice because the numbers are not "random" enough and that the probability of these numbers coming up is less than the probability of six "random" numbers being chosen. Is John right? Explain your answer.

7.2 Keno

Keno is a game with lottery overtones (and is, in fact, played as a lottery in a number of states), but there are some differences, chief among them being that the player picks the number of numbers that he or she will select. It has been played at casinos for years but is probably not as well-known as other games like roulette or poker. Various casinos have added some wrinkles to the game, but in this section, we will focus on the most basic version of Keno and look at a few very simple calculations.

The basic game is played as follows: the player must select some numbers that are between 1 and 80. The player decides how many numbers to select; if, for example, she decides to pick six numbers, then this is a "pick 6" game (and likewise, of course, if she selects some other number). The casino then selects 20 numbers from 1 to 80; this can be done live (using a ball-bouncing machine) or (more likely these days) electronically. The amount the player wins depends, of course, on how many of her numbers show up in the 20 "good" numbers selected by the casino. The payoff amount depends on the casino and the number of numbers that the player has picked.

Example 7.2.1. What is the probability that, in a pick 6 game, the player matches all six numbers?

Solution. As always, the answer is a fraction: the denominator is the total number of ways that one can pick six numbers in the range from 1 to 80. This is, of course, $\binom{80}{6} = 300{,}500{,}200$. The numerator is the total number of "favorable" outcomes, i.e., the total number of ways of selecting

Lotteries and Keno 73

six numbers from the 20 "good numbers" selected by the casino. This is $\binom{20}{6} = 38{,}760$. Dividing 38,760 by 300,500,200 gives (approximately) .00013, the desired (and very small) probability.

Of course, to win at a casino, the player need not necessarily match all her selected numbers. With this in mind, let us modify the previous example.

Example 7.2.2. What is the probability that, in a pick 6 game, the player matches exactly five of her numbers?

Solution. The denominator is the same as in the previous example. The numerator, however, is different. Here we must first select five "good" numbers and one "bad" one. The five good numbers can be selected in $\binom{20}{5} = 15{,}504$ ways. Then the 6th "bad" number must be selected from the 60 numbers not selected by the casino; this, of course, can be done in $\binom{60}{1} = 60$ ways. The product of these two numbers, 930,240, is the total number of favorable outcomes, and so the probability asked for in this example is $930240/300{,}500{,}200$ or, approximately .003.

To determine the expected value of a Keno bet, we need to know the payoffs made by a casino. Since there will typically be more than one payoff depending on how many matches one makes, computing the expected value of, say, a one-dollar bet can be a fairly cumbersome undertaking. A book that does the calculation in the context of a "pick 10" bet is Packel's book *The Mathematics of Games and Gambling* [Pa]; we will summarize his calculations here. First, Packel gives a payoff schedule:

Number Matched	Payoff
10	$10,000
9	$2,600
8	$1,300
7	$180
6	$18
5	$2

Matching less than five numbers results in no payment by the casino at all. From the definition of expected value, it follows that the expected value of a $1 "pick 10" bet is

10,000 P(match 10) +

2600 P(match exactly 9) +

1300 P(match exactly 8) +

180 P(match exactly 7) +

18 P(match exactly 6) +

2 P(match exactly 5) +

−1

The −1 term at the end reflects the fact that the player has already spent $1 to play, so the player's actual expectation is one dollar less than what is received from the casino. (If the casino gives the player two dollars, for example, the player only makes a one-dollar profit.) We could have avoided this term by just subtracting 1 from each outcome from 10,000 to 2 and then changing the last line to read " −1 P(match less than 5)", but doing it this way seems easier.

The individual probabilities that appear in the sum above can be calculated just as in examples 7.2.1 and 7.2.2 above, multiplying and dividing binomial coefficients. For example, the probability that one matches exactly 8 out of 10 winning numbers is $\binom{20}{8}\binom{60}{2}/\binom{80}{10}$. Using a binomial coefficient calculator webpage, one can specify the precise numerical values of the three binomial coefficients that appear in this expression; one can then use a calculator to do the division. The precise value, Packel tells us, is .0001354194. By computing all the specific probabilities this way and then doing the addition indicated above, Packel arrives at the answer −.20742, meaning that the gambler loses almost 21 cents on each dollar bet. While not as bad as the numbers racket or "pick 6" lottery, this is still a terrible bet, roughly four times worse, say, than roulette.

Exercises

7.9. According to the Ohio Lottery webpage, that state offers a Keno lottery game in which the player can pick from 1 to 10 numbers from the numbers 1 to 80. In a "pick 1" game, the player wins $2 (on a $1 bet) if the one number picked is one of the 20 lucky numbers picked by the lottery. What is the expected value to the player of this bet?

7.10. In a "pick 2" game in the Ohio Lottery, the player must "catch 2" in order to win anything at all, in which case he or she wins $11 (i.e., a $10 profit). What is the expected value to the player of this bet?

7.11. The Ohio Lottery, in its "pick 3" game, pays 2:1 if the player catches 2 and 27:1 if the player catches 3. What is the probability of the player winning any prize at all?

7.12. (Continuation of previous exercise.) What is the expected value of a "pick 3" Keno game?

8

Blackjack

In this chapter, we take a look at Blackjack, another mainstay of virtually all American casinos, and also one of the fairer (i.e., least unfair) games offered. It is quite difficult to completely analyze mathematically because the probabilities change constantly as the game is played, so we will content ourselves with a brief introduction.

8.1 Rules of the Game

The rules of blackjack that we will set forth here are fairly common but they are not entirely standard and can vary from casino to casino. The underlying idea, however, is that the gambler is dealt cards, each with a point value. The gambler, who decides when he or she wishes to stop being dealt cards, wants to get as close to a total of 21 without going over, and, in particular, wants to get closer to 21 than the dealer.

More specifically: a numbered card (2, 3, ..., 10) is assigned its number as its point value; a face card (King, Queen, and Jack) is assigned point value 10, and an Ace is assigned point value 1 or 11 at the player's discretion (and the player can change the value as the game proceeds).

Several players sit at the table with the dealer, who deals all players and herself two cards (the players' cards are usually dealt face up; for the dealer, one card is face-up and the other is face-down). The first player can then request more cards ("hit me") as many times as he or she wants until that player either goes over 21 ("busts") and loses (regardless of what the dealer ultimately winds up with) or reaches a point sufficiently close to 21 that he or she chooses to stop ("stand"). The same ritual then takes place with all the other players at the table. After each player has either busted or stopped at a number less than or equal to 21, the dealer then turns over her down card and (perhaps) deals herself some more. According to the rules in most American casinos, the dealer must hit herself if she has 16 or less, and must stand on 17 or more.

If a player has not busted and the dealer gets closer to 21 than he or she does, the casino wins that player's bet. If the player got closer to 21 than the dealer did, he or she wins (even money), except: if the player gets exactly 21 on the first two cards, that's a "blackjack" (or a "natural"), and if the dealer does not also get a blackjack, the player wins 3:2 (i.e., he or she makes a $3 profit on a $2 bet). If the player and the dealer both wind up with the same number of points (including if they both get blackjack), that's a "push", i.e., a tie: no money changes hands.

These rules are a bit oversimplified in that we have ignored various things that a player can, in certain circumstances, do, such as splitting pairs or doubling down. There is another kind of bet that a player can make that we will discuss, however, and that is an *insurance* bet; discussion of that will appear in the next section.

You might be wondering at this point how this game favors the casino since the rules seem pretty symmetrical: there is no reason, after all, to think that the dealer can get closer to 21 than a player can. Indeed, the player even has an advantage that the dealer does not: the player can make decisions regarding whether to be hit or not based on his or her observation of the cards that are already visible, whereas under the rules as stated above, the dealer cannot exercise any discretion. However, the casino does have one major advantage, which accounts for the house edge: if a player busts, he or she loses automatically, regardless of what happens with the dealer. If the player is the only one at the table, the dealer need not even deal herself any cards at all. If the dealer does deal herself some cards (because of other players at the table) and winds up busting, the player who busted first still loses.

Another point to briefly mention here (and which will be elaborated on later in the chapter) is that the dealer may not be dealing from a single deck; sometimes multiple decks are put together and shuffled together, and cards are drawn from this "multi-deck".

Exercises

8.1. Three players are at a table with the dealer. The dealer's up card is a 9. Player 1 is dealt two Kings and chooses to stand. Player 2 is dealt a 10 and a 2, asks to be hit, and gets a 6, at which point he stands. Player 3 is dealt a 6 and a 7, asks to be hit, and gets a Jack. The dealer then exposes her down card, which is an 8; pursuant to house rules, she stands at 17. Among the three players, who has won?

8.2. You are the first of eight players at a table. You are dealt a 7 and a 6. Player 2 is dealt two 3s. Player 3 is dealt a 2 and a 4. Player 4 is dealt a 3 and a 5. Player 5 is dealt a 2 and a 3. Player 6 is dealt a 5 and a 6. Player 7 is dealt an Ace and a 2. Player 8 is dealt a 5 and a 4. The dealer's up card is a 7. Ordinarily, since you have a 13, you might be inclined to be hit, but in this case, the cards that have already been dealt might give you pause. Why?

Blackjack

8.3. You have been dealt a 10 and a 5 by the dealer. The player next to you has an 8 and a 9, and the third and last player at the table has a King and an 8. The dealer's up card is a 7. Should you hit or not? Give a reasonable argument in support of your answer.

8.4. Two players sit at a table with the dealer. The first player is dealt a 10 and a 2, asks to be hit, and gets a Jack, thereby busting. The second player gets a 9 and a King, and stands. The dealer's up card is a 10 and her down card a 6. Under the rules, as we have stated, the dealer must hit herself. What is the probability that she beats player #2?

8.2 Basic Blackjack Calculations

In this section, we do some basic probability calculations involving blackjack. It will be recalled that at the end of the previous section, it was noted that the dealer may not be dealing from a single deck, but in these calculations, unless otherwise specified, it will be assumed that he or she is. We start with the question of how likely it is to be dealt a blackjack on the first two cards.

Example 8.2.1. What is the probability that a player, dealt two cards from an unused deck, will be dealt a blackjack?

Solution. The total number of ways of being dealt two cards from 52 (with order not counting) is $\binom{52}{2} = 1,326$. This will, of course, be our denominator in the fraction that we are about to compute. In order to be dealt a blackjack, the player must receive a 10-point card and an 11-point card. There are 16 10-point cards (12 face cards and four 10s), so the total number of ways of selecting one is 16. Likewise, the only 11-point cards are the aces, of which there are four in a single deck. So, there are four ways to be dealt a single 11-point card. Therefore, the total number of ways of being dealt two cards totaling 21 points is 64. Dividing this by 1,326 gives .048. So, the likelihood of being dealt a natural is roughly 1 in 20. Of course, if other players' cards are visible, that will affect the answer.

Another simple kind of probability question involves the question of whether a player should hit or stand.

Example 8.2.2. Five players are at a table with a dealer. Player 1 is dealt a King and a 5. Player 2 is dealt two Jacks. Player 3 is dealt a 3 and a 6. Player 4 has an Ace and an 8, and Player 5 has an 8 and a 9. The dealer's up card is an Ace. If Player 1 asks to be hit, what is the probability that he will bust?

Solution. Player 1 has a total of 15, so, if hit, he will bust if he receives a card with point value 7, 8, 9, or 10. There are 11 cards showing, so there are 41 possibilities for the next card to be dealt. There are four 7s, two 8s, three 9s, and 13 10-value cards left in the deck, so the total number of cards that will bust Player 1 is 22. Thus, the probability that Player 1 will bust is $22/41$, a little more than half.

There is a side bet in blackjack called *insurance*, which you can take when and only when the dealer has an Ace showing. In effect, you take out another bet, insuring against the probability that the dealer already has a blackjack. You may bet up to one-half the amount of your initial bet, and you win at 2:1 if the dealer has blackjack. Most blackjack strategy books advise against taking an insurance bet unless you are very adept at keeping track of cards, though it is possible (see the exercises) to have a positive expected value on an insurance bet. Here is an example where the expected value is negative:

Example 8.2.3. Suppose the dealer has an Ace showing, and you, the only player, have a King and a Queen. Let's say you have an insurance bet of $1. This means that, on this insurance bet, you will win $2 (outcome +2) if the dealer has a blackjack and will lose the bet (outcome –1) if not. What is the expected value of this bet?

Solution. There are 49 cards left in the deck. Of the 16 cards that could give the dealer a blackjack, you already hold two of them; therefore, there are 14 left. So, the probability that the dealer has a blackjack is $14/49$. The expected value on your insurance bet, therefore, is $2(14/49) - (35/49) = -7/43$ or roughly –.143. An insurance bet under these circumstances, therefore, loses more than 14 cents on the dollar on average.

Another interesting question that one might ask is: what is the probability of winding up with a certain number of points? If you follow the dealer's standard rule (hit on 16 or less, stop on 17), then clearly you will never wind up with less than 17 points, so the possible outcomes here are 17, 18, 19, 20, 21 (in three or more cards), blackjack, or bust. We have already determined that the probability of a blackjack is about 1/20 or .05. The remaining probabilities are extremely difficult to calculate because there are so many ways that each outcome can be obtained, but computers have been used to determine these probabilities. This list of probabilities appears in a book [Olo] by Olofsson called *Probabilities: The Little Numbers that Rule Our Lives*. In the chart below, the number on the left is the outcome, and the number on the right is the probability, rounded off:

17	.15
18	.15
19	.14
20	.18
21	.05
Blackjack	.05
Bust	.28

Blackjack

Using these figures, we can do some interesting probability computations. For example:

Example 8.2.4. If you get to a 17 on your cards, what is the probability you will win?

Solution. Because the dealer must hit on a 16, he or she must always wind up with at least a 17, or bust. Therefore, the only way that you can win with 17 is if the dealer busts, which (from the chart above) happens with probability .28. With a total of 17 in your hand, you will tie with a probability of .15, and lose if the dealer gets 18, 19, 20, 21, or blackjack, which (adding the probabilities) gives .57.

Now, suppose you have gotten to 18 in your hand. You will win if the dealer winds up with 17, or busts. As a homework problem, you are asked to use this observation to determine the probabilities that you will win, lose, or tie with an 18.

Exercises

8.5. Redo example 8.2.1, this time under the assumption that the dealer is using two decks.

8.6. In a single-deck situation with no other cards visible, determine the probability that a player is initially dealt two cards, adding up to 20 points. (Begin by thinking of the various ways this can happen.)

8.7. The "Royal Hand" consists of a King and Queen of the same suit. Compute the probability of being dealt a Royal Hand in the first two cards dealt from a single deck.

8.8. Refer to the situation described in exercise 8.2. Again, assuming a single deck, what is the probability that if you ask to be hit, you will bust?

8.9. Suppose that in example 8.2.2, Player 1 asks to be hit and is dealt an Ace. What is the probability that, if hit again, Player 1 will bust?

8.10. Three players are at a table with a dealer. Player 1 has a King and an Ace, Player 2 has a 7 and an 8, and Player 3 has a Queen and a Jack. The dealer's up card is a Queen. At this point in time, what is the probability that Player 1 will win?

8.11. Three players are at a table. The dealer's up card is an ace. If each of the three players takes an insurance bet, does the expected value of the bet change with the player, or is it the same for all three? Explain.

8.12. Suppose the dealer has an Ace showing, you have a King and an 8 in your hand, and another player at the table is showing a 5 and a 6. Compute the expected value of a one-dollar insurance bet placed by you.

8.3 Card Counting

We can't talk about blackjack without at least mentioning card counting. It seems safe to say that anybody who has heard of blackjack has probably at least heard of card counting, even without, perhaps, knowing any of the details of this method of play. The idea of card counting seems to date back to the early 1960s and the publication of a book [Tho] entitled, *Beat the Dealer*, by a mathematician named Edward Thorp. There are a number of different versions of card counting; we will briefly discuss the easiest of them, called High-Low.

This method is based on the idea that a deck that is rich in high-value cards like 10s and Aces tends to favor the player rather than the dealer. There are several reasons for this, including (a) a deck with lots of Aces and 10-value cards can produce more blackjacks, which favor the gambler rather than the casino because the gambler wins more than the amount of the bet on a blackjack whereas the casino only wins the amount of the original bet if the dealer gets one; (b) a deck with a lot of 10-value cards in it makes it more advantageous for the player to take insurance bets; and (c) because the dealer is obligated to hit on a 16, a deck with high-value cards increases the possibility that he or she will go bust. While it is not practicable to actually memorize each card that has been exposed as the game proceeds, the High-Low method provides a way of quickly keeping track of whether the cards remaining in the deck have large or small value. The method works as follows: as the cards are exposed, give a value of +1 to every card with value 2 through 6, a value of 0 to every card with value of 7, 8, or 9, and a value of −1 to every card with value 10 or 11. If you keep a running tally going, then a positive value indicates the presence of cards of value 10 or 11; that's a "good" deck and might justify more heavy betting.

Card counting, if done very well, can actually result in the casino's expected value becoming slightly negative, giving the player an edge. So, how did casinos react to this new development? Card counting (if done without the aid of electronic devices, at least) is legal; after all, one could hardly pass a law prohibiting a player from playing intelligently. However, even though card counting is legal, casinos have the right to require any gambler to leave for any non-discriminatory reason. So, if casino officials suspect a person of card counting, they can simply order that person to leave. A person who refuses to leave after being told to do so can then be arrested for trespassing. Another way that casinos can deal with card counters is to use multiple decks and shuffle frequently, thereby making it more difficult to keep count of what is left.

Card counting has appeared in popular media as well as technical mathematics books. For example, Ben Mezrich wrote a book called *Bringing Down the House*, about six math geniuses at MIT who took casinos for a fortune. (The Boston Globe later wrote an article saying that this book contained

Blackjack

significant fictional aspects.) The book was in turn made into a movie, *21*, starring Kevin Spacey, which even mentioned the Monty Hall problem, previously discussed in Chapter 3 of this book.

Exercises

8.13. Five players are at a table with a dealer in a single-deck game. The players have been dealt, respectively, a 3 and 7; King and 9; 2 and 5; and 4 and 6. The dealer's up card is a 5. What is the High-Low count? Does the remaining deck favor the dealer or the players?

8.14. You are playing single-deck blackjack at a table with many other players. You can observe that all four aces are visible. The High-Low count is -5. You have a 16. Do you think you should ask to be hit? Explain.

9
Farkle

Any person with a Facebook account has perhaps seen some reference to the game of Farkle, a dice game with poker overtones. In this chapter, we look at some of the mathematics and probability theory behind this game. Because the game has, literally, an infinite number of possible outcomes, a complete analysis is perhaps impossible, but the work we have done to date at least allows us to look at some interesting questions regarding the game. We first discuss the rules of the game, at least as it is played on Facebook.

9.1 Rules of the Game

The rules of the game are not difficult. Basically, the goal is to throw dice and try to accumulate as many points as possible, with these points being awarded for certain combinations that resemble poker hands. There are apparently some variations in scoring elsewhere, but our discussion will be based on the Facebook rules. Also, in the interests of keeping the analysis simple, we ignore certain Facebook variations using "bonus points" that can be earned and used for various benefits, such as doubling the point value on a given round.

The game consists of ten rounds. Any given round commences with the player (or the player's computer) rolling six dice. Point values are assigned to some individual die and combinations of dice, as follows:

- A single die showing 1 counts 100 points, and a single die showing 5 counts 50. These are the only individual die that have point value.
- A "straight" (six dice, showing each of the numbers 1 through 6), counts 1500 points.
- Three pairs (e.g., two 2s, two 3s, and two 4s) count 750 points.
- Three 1s count for 1000 points, with 1000 added for every additional 1. Three 2s count for 200 points, three 3s count for 300 points, three 4s count for 400 points, three 5s count for 500 points, and three 6s count for 600 points, with, again, the same amount added for four of a kind, added again for five of a kind, and added yet again for six of a kind. Thus, for example, six 1s count for 4000 points; four 2s count for 400 points; five 2s count for 600 points, etc.

Farkle

Here are some examples: if you roll 1, 2, 3, 3, 4, 6, you have obtained only 100 points, but if you roll 1, 3, 3, 3, 4, 6, you have obtained 400. (Make sure you understand why both of these statements are true.) Likewise, rolling 1, 2, 2, 2, 2, 2 gives you 700 points. As a check on your understanding, determine the point value of each of the following combinations of dice: 2, 2, 3, 4, 4, 6; 1, 1, 3, 4,5, 6; 1, 2, 2, 2, 6, 6; 1, 1, 1, 1, 5, 6. (Answers: 0; 250; 300; 2050.)

After the computer has performed its initial roll of six dice, the player gets to exercise some strategy: he or she can retain or "bank" any one or more dice that has some positive (i.e., nonzero) point value and roll the remaining die. For example, if the original roll yielded 1, 2, 3, 4, 4, 5, the player can bank the 1 and have the computer roll the remaining five dice, or bank the 1 and 5 and have the computer roll the remaining four. If at any point the roll of the dice results in a combination worth no points, the player has "farkled" and the round ends with 0 points for that round, losing any points that may have already been earned. (On Facebook, when a player farkles, the computer makes a very unpleasant noise, unless the player has had the sense to mute the volume before starting the game.) Once the player has retained some dice, the remaining die can be re-rolled, and the process begins anew. This process of retaining and re-rolling can continue indefinitely. The round ends when either the player farkles, or, once the player stops the round and banks all the points obtained to that point. The player can only do this, however, when 300 or more points have been accumulated.

For example, suppose the initial roll of the dice yields 1, 3, 3, 5, 6, 6. The player banks the 1. Since only 100 points have been accumulated, the player must roll the remaining die. Suppose the computer rolls 1, 1, 2, 3, 4. The player can bank either one or both of the 1s. If only one 1 is banked, the remaining four dice are rolled. If both 1s are banked, there are now 300 points accumulated (note that the 1 that was initially banked is NOT combined with these 1s; each roll is judged separately from all the others), and the player has the option of closing the round with a total of 300 points or rolling the remaining three dice. This can result in a good outcome (all three dice coming up 1s, for example) or a bad one (the player rolls 2, 2, 3, which results in a farkle and a total of 0 points for this round). If all three dice were to come up 1, the player would have earned 1,300, and since the dice have been used up, the computer will continue the round by rolling six new dice. And this is where strategy enters the picture: should the player take the 300 points for the round and stop, or gamble on doing much better, knowing that there is the risk of farkling and losing everything?

Let us close this section by recounting the details of an actual farkle round played by the author online. The initial six dice showed 1, 2, 4, 4, 5, 6. The 1 was retained, and five dice were re-rolled, giving 1, 3, 4, 5, 6. Again, only 1 was retained, and four dice were re-rolled. The computer rolled 1, 1, 3, 5. Being risk-averse, the author retained 1, 1, and 5 and closed the round with a score of 450. Had the 3 in the final throw been a 5 (or, even better, a 1), then all the dice could have been banked and the round continued with six fresh new dice.

Exercises

9.1. Your first roll of the dice shows 1, 2, 3, 4, 4, 5. What are all possible legal moves at this point? Of these possible moves, which one is the most illogical? Explain.

9.2. Give a specific example of farkling with one die, two dice, three dice, four dice, five dice, and six dice.

9.3. Give two examples of a six-dice farkle hand worth exactly 1500 points.

9.4. What is the maximum number of points you can earn on a single roll of six dice? Give an example of a way you can do this.

9.5. Your first roll of the dice shows 1, 1, 1, 1, 3, 5. How many points is this hand worth? What would your next move be if you rolled this hand? Why?

9.6. Is it possible to score exactly 750 points on a roll of six dice that contains exactly three 5s in it? Explain.

9.2 Some Probability Calculations

In Section 9.3, we will discuss what the probability of farkling is when one die is rolled, when two dice are rolled, etc. First, though, let's consider some other typical events that may occur in a game of Farkle and determine their probabilities.

Example 9.2.1. What is the probability of rolling a 1500-point "straight" on the initial roll of the dice?

Solution. It is convenient to think of this as an "order counts" situation (think of the six dice as being differently colored, for example). In that case, the total number of ways of rolling six dice is $6^6 = 46{,}656$, since each of the six dice can show any of six numbers. This is our denominator. Now, for the numerator: how many of these outcomes yield the numbers 1 through 6 in some order? Clearly, this possible combination of six different numbers can arise in $6! = 6 \times 5 \times 4 \times 3 \times 2$ ways. So, the desired probability is $720/46{,}656$. This number is roughly .0154, or (rounding) about 3/200.

Example 9.2.2. What is the probability of rolling all 1s on the initial roll of the dice?

Solution. As before, if we think of the dice as distinguishable, then our denominator is 6^6. This time, however, our numerator is 1; there is only one way for each die to show a 1. So, the probability of rolling all 1s is $1/46{,}656$.

Farkle

Example 9.2.3. What is the probability of rolling all 2s on the initial roll of the dice?

Solution. There really isn't any difference between this example and the one before it, is there? There's still only one way each die can show a 2. So, the probability here is also $1/46{,}656$. Note, however, that this example shows an interesting thing: the point value of a configuration of dice does not entirely depend on the probability of that configuration being rolled. "All 1s" and "All 2s" are equally likely events, yet have dramatically different point values.

Example 9.2.4. What is the probability of getting at least one 1 in the first roll of six dice?

Solution. If you try and count the number of ways in which at least one 1 shows up, you will quickly realize this is an incredibly complicated problem, especially since there can be one 1, two 1s, etc. However, remember that it is sometimes advantageous to compute the probability of the complementary event (which, in this case, is "no 1s show up") and then subtract from 1. In this case, the probability of the complementary event is very easy to compute: for there to be no 1s at all, every die must be one of five possible outcomes (i.e., 2, 3, 4, 5, or 6), so the total number of ways this can occur is 5^6. The probability of no 1s, therefore, is $(5/6)^6$. It follows that the probability of at least one 1 is $1 - (5/6)^6$.

Example 9.2.5. What is the probability of getting three pairs on the first roll of six dice?

Solution. You may wish to first review Section 6.3, where the probability of getting two pairs in poker was discussed. If you recall, the first step there was to pick the two ranks that would be that pair. So, by analogy, we now pick the three numbers that will constitute our pairs, which we can do in $\binom{6}{3} = 20$ ways. Having picked our three numbers (let's say we picked 2, 4, and 5), we then have to arrange them in six slots. There are $\binom{6}{2} = 15$ ways to pick the two slots in which we will put the 2s. Then, from the four slots remaining, there are $\binom{4}{2} = 6$ ways to pick the two slots in which to put the 4s, and then there is only one choice left in which to put the 5s. So, our numerator is $20 \times 15 \times 6$, or 1800, and the probability of getting three pairs is thus $1800/46{,}656$.

Example 9.2.6. On the initial roll, what is the probability of rolling exactly 4 1s?

Solution. This probability is a fraction, the numerator of which is the total number of ways of rolling exactly four 1s. There are $\binom{6}{4} = 15$ ways to pick the four slots that will be occupied by 1s. Each of the two remaining slots can be filled in five ways (anything but a 1 in each slot). So, our numerator is $15 \times 25 = 375$, and the probability of getting exactly four ones is $375/6^6$.

Exercises

9.7. On a roll of six dice, what is the probability of rolling three pairs?

9.8. Consider the situation described in exercise 9.5. If you retain the four 1s and roll two dice, what is the probability that you will improve the value of the hand that you already have?

9.9. You roll a single die. What is the expected number of points you will get on that roll? (Ignore any points that you may have already banked.)

9.10. What is the probability of earning exactly 250 points on a roll of three dice?

9.11. What is the probability of earning exactly 200 points on a roll of three dice? Why is this problem more difficult than the preceding one?

9.3 Should You Risk Another Roll? The Probability of Farkling

In making strategic decisions about whether to bank what we have and terminate the round or proceed and try to get more points (at the risk of farkling and losing everything), it is, of course, helpful to know what the probability of farkling is. One preliminary observation is pretty obvious: the more dice you roll, the less chance you have of farkling, simply because with more dice you have a better chance of coming across a 1 or 5 or some other combination that yields points. So, if you already have a large number of points accumulated for a single round, you would be well advised to stop rather than roll one die, but you are probably pretty safe if you can roll five. But what is the exact probability of farkling? We next turn to this question, computing the probability of farkling as a function of the number of dice to be rolled.

One Die. If a single die is rolled, there are, of course, six possible outcomes, four of which (2, 3, 4, 6) will result in farkling. So, the probability of farkling in this case is $2/3$.

Two Dice. If two dice are rolled, there are 36 possible outcomes, 16 of which result in a farkle. So, the probability of farkling here is $16/36$. This is approximately .444, so even with only two dice rolled, it is more likely than not that a farkle will not occur.

Three Dice. When you roll three dice, there are 216 possible outcomes. In order to farkle, each roll can come up in four ways (just as explained in the previous paragraphs). However, even if every roll comes up a 2, 3, 4, or 6, you can still avoid farkling if any of these numbers come up all three times, a circumstance that can happen four ways (all three 2, all three 3, etc.). So, the total number of ways that will produce a farkle is $4^3 - 4 = 60$ and the probability of farkling is, therefore, $60/216$, which is approximately .2778, not quite three times in ten.

Four Dice. When you roll four dice, there are $6^4 = 1296$ possible outcomes. How many of these yield no point values at all? First of all, for this to happen, each of the four dice must show a 2, 3, 4, or 6, so there are at most $4^4 = 256$ possible outcomes. But even if all the dice show one of these four numbers, we can still get some points if there is a three of a kind, or a four of a kind, so we have to subtract from 256 the number of three, or four, of a kind that can appear using these numbers. There are only four ways to get four of a kind (all four 2, all four 3, etc.). As for three of a kind, there are $\binom{4}{3} = 4$ ways to pick the three dice out of four that will show this repeated number, and four choices to pick the number that will be repeated. Then there are three choices for the remaining die. So, the total number of three-of-a-kind rolls is $4 \times 4 \times 3 = 48$. Thus, the total number of ways to farkle with four dice is $256 - 48 - 4 = 204$, and so the probability of farkling is $204/1296$, which is about .1574.

Five Dice. We could mimic the idea used in the previous paragraph, but then we would have to subtract the number of various possible ways to get three of a kind, four of a kind, or five of a kind, and this would get rather cumbersome. So, we will directly count the number of ways to roll five dice and farkle.

Since every die must show one of four numbers, and there are five dice, some numbers will have to be repeated; on the other hand, no number can repeat more than twice. So, we have two possibilities here: one number is repeated twice, and the other three dice show different numbers (call this pattern XXYZW), or two different numbers appear twice, and the fifth die shows a third number (pattern XXYYZ).

Let us first count the number of ways pattern XXYZW can occur. We can pick the repeated number X in four ways and find two slots out of five for it in $\binom{5}{2} = 10$ ways. We can fill the three remaining slots in $3! = 6$ ways. Thus, the first possibility can occur in $4 \times 10 \times 6 = 240$ ways.

As for the second pattern, there are $\binom{4}{2} = 6$ ways to select the two repeated numbers. The first of these can be put into two slots out of five in $\binom{5}{2} = 10$ ways. The second of these can be put in two of the three remaining slots in $\binom{3}{2} = 3$ ways. Finally, there are two numbers left to put in the fifth slot. So, all told, this pattern can occur in $6 \times 10 \times 3 \times 2 = 360$ ways.

Thus, there are a total of $360 + 240 = 600$ ways to farkle. To determine the probability of farkling, therefore, we divide this number by the total number of ways of rolling five dice, which is 6^6. The result is approximately .0771. So, when rolling five dice, there is less than a one in ten chance of farkling.

Six Dice. When six dice are rolled, in order to farkle, no one number can be repeated more than twice, and we can't have three doubles. Since there are only four possible numbers: two numbers must appear twice and the two remaining numbers must appear once. Call this pattern XXYYZW, where X, Y, Z, and W are the four numbers 2, 3, 4, and 6.

There are $\binom{4}{2} = 6$ ways to select the two numbers (X and Y) that will appear twice. There are then $\binom{6}{2} = 15$ ways to select two slots to put that number in. We can then select $\binom{4}{2} = 6$ ways to select two slots out of the remaining four in which to put the second of our "repeating" numbers. We now have two numbers (Z and W) to put in two slots, which can be done in $2! = 2$ ways. So, the total number of ways to farkle by rolling six dice is $6 \times 15 \times 6 \times 2$, or 1080. The probability of farkling, when six dice are rolled, is therefore $1080/6^6$, which is approximately .023. In other words, we can, in the long run, expect to farkle less than three times out of 100 when rolling six dice.

Exercises

9.12. Consider the situation described in exercise 9.5. If you retain the four 1s and roll two dice, are you more likely to improve the value of your hand, or to farkle?

9.13. If you roll three dice, what is the probability that two threes will appear, and you will also farkle?

9.14. If you roll three dice, what is the probability that three different numbers will appear, and you will also farkle?

9.15. If you roll four dice, what is the probability that four different numbers will appear, and you will also farkle?

10

An Introduction to Game Theory

The games that we have looked at so far in this book can be divided into two categories. In some games, like craps or roulette, once the initial decision as to the bet itself is made, the player is not called upon to make any more decisions and just lets nature take its course. But other games, like poker and blackjsck, require the player to make strategic choices as the game progresses: should I ask to be hit? Which cards from my poker hand should I discard? Should I bluff or fold?

Strategic decisions are, of course, not unique to casino games or other games of chance. They come up in numerous situations in real life as well—for example, in business or war. Precisely because of the prevalence of decision-making in uncertain situations, a branch of mathematics known as game theory came to be developed. In this chapter, we give a very elementary introduction to this area of mathematics. Game theory is a huge subject on which entire books can (and have) been written; we look here only at the tip of the game theory iceberg.

10.1 Introduction and Basic Definitions

We begin our glimpse of this subject by looking at some basic terminology. In particular, we try and give some sense of what the word "game" means in the context of "game theory". We will see that in some ways, a mathematical "game" is broader and, in some ways, narrower than a "game" in the ordinary, intuitive sense.

For example, the children's game Candy Land, though likely considered a "game" in ordinary English parlance, is not a "game" in the sense of this chapter, simply because it involves no strategy at all: a player draws a card with a color on it, and movies his or her piece to the next square of that color. There are no decisions to be made by a player. Likewise, the card game War, in which two players turn over a card from a stack, with each person hoping to have the high card, is not a mathematical "game" for precisely the same reason: there is no strategy involved.

On the other hand, there are activities, such as military battles, that are not in the least bit humorous, entertaining, or fun, that can be analyzed as mathematical "games" because they involve strategic choices to be made by two or more people. For example, in the next section of this book, we will look at The Battle of the Bismarck Sea, an interesting example of a mathematical game—and one that actually took place. Also, the Cuban Missile Crisis, more than half a century ago, could be considered a "game", even though it was deadly serious, with literally the fate of the world hanging in the balance.

A game between two players is called *zero-sum* if one player loses what the other wins. A typical gambling game is zero-sum; for example, if a gambler loses $5 at blackjack, that's what the casino (or opposing player) wins. There is no room for cooperation between the players in a zero-sum game; their interests are diametrically opposed to one another.

There are non-zero-sum games, and they can be very important, but in this book, in the interest of keeping things as simple as possible, we will only consider zero-sum games. Also, for the sake of simplicity, we will only consider two-person games.

Here is a simple, if somewhat artificial, example of a two-person zero-sum game: two players, Rose (R) and Colin (C), each simultaneously display a nickel or a dime. If the two coins match (both nickels or both dimes), then Rose takes both coins; otherwise, Colin takes both. It is clear that this is a zero-sum game since one player wins what the other loses. Each person has two strategies (display a nickel or display a dime), and there are a total of four possible results:

Both show a dime; Rose wins 10 cents

Both show a nickel; Rose wins 5 cents

Rose shows a dime, Colin shows a nickel; Rose loses 10 cents

Rose shows a nickel, Colin shows a dime; Rose loses 5 cents

We can conveniently summarize this information in a *matrix*, or array of numbers, where each row corresponds to a strategy choice for Rose, each column corresponds to a strategy choice for Colin (now you see why the players' names begin with R and C), and the intersection of any row and column shows the outcome from Rose's (the row player's) perspective. The matrix, with strategies labeled, is:

	Colin: nickel	Colin: dime
Rose: nickel	5	–5
Rose: dime	–10	10

Using this array, we can make some remarks. All of them are very straightforward but do at least point in the direction of more interesting observations. For one thing, if Rose was a greedy type who wanted to get the most that she possibly could, she would select showing a dime as her strategy. But if Colin anticipated this strategic choice, he would select showing a nickel as his choice, and Rose would lose the most she could possibly lose. Likewise, if Colin wanted to get the most he could, he would select showing a nickel as his strategy (remember the payoffs are from Rose's perspective, so Colin wants the payoff with the smallest number, unlike Rose, who wants the payoff with the largest one). But if Rose anticipates Colin's choice, she will show a nickel, and wind up winning five cents. So, selecting a strategy that offers the possibility of a maximum win offers some disadvantages.

There is an alternative approach for Rose and Colin, one that may be summed up by the phrase "hope for the best, prepare for the worst". Rose might ask herself: what is the worst that can happen to me for each possible strategy? If she selects a nickel, she can, at worst, lose a nickel; if she selects a dime, she can, at worst, lose a dime. So, if she were the cautious type who wanted to ensure that she loses as little as possible, she would select a nickel. We might call this her "maximin" strategy because in each row she first identifies the smallest number (the worst possible outcome for her) and then selects the largest of these numbers (the best of the worst outcomes). Another name for this strategy is the *prudent* strategy.

Likewise, Colin (for whom big numbers are bad) would, if adopting a similar strategy, reason that if he selects column 1 (show a nickel), the worst that can happen to him is that he loses a nickel; if he selects column 2 strategy (show a dime), the worst that can happen is that he loses a dime. So, the best of these bad outcomes (his "minimax" strategy) is that he loses a nickel. That would be his prudent strategy.

It is interesting to note that US military commanders are apparently required to use the prudent strategy in making command decisions. A 1954 paper by a former Air Force officer named O.G. Haywood titled *Military Decision and Game Theory*, appearing in Volume 2, Number 4 of the Journal of the Operations Research Society of America, refers to the US "doctrine of decision" that "a commander base his actions on his estimate of what the enemy is capable of doing to oppose him."

It should be clear that, in this particular game, if either player knows in advance what strategy the other will follow, that information can be used to the player's advantage. In particular, if this game were to be played repeatedly, it would not be to any one player's advantage to follow one particular strategy all the time—the other player would quickly discern this, and select his or her counter-strategy accordingly. What a player should do is select each strategy a certain percentage of the time (using, perhaps, some randomizing device to determine which strategy to select in any given play of the game). Interestingly, this is not always the case with two-person zero-sum games; we will soon see examples of such games where knowledge of the other person's choice of strategy should have no bearing on your strategic choice.

Exercises

10.1. Suppose that Rose and Colin play the familiar game of "rock paper scissors", where each person simultaneously shows a flat hand (paper), their index and middle fingers (scissors), or a fist (rock). In this game, scissors beat paper, paper beats a rock, and a rock beats scissors. Assume the winner of any round wins $1 from the other player. Model this as a zero-sum game matrix. Proceeding purely intuitively, without attempting to justify your answer mathematically, what do you think is a good long-term playing strategy for either player? (Interestingly, James Bond was presented with this issue in *You Only Live Twice*.)

10.2. Consider a zero-sum game where Rose has two strategies R1 and R2, Colin has two strategies C1 and C2, and the game matrix is given below. Explain why Rose really has only one rational strategy, and likewise for Colin. Explain why, even if Rose knows what Colin is going to do, her strategy choice is not affected.

	C1	C2
R1	6	0
R2	-1	-5

10.2 Zero-sum Games: Domination

In this section, we take a look at the concept of "domination" in zero-sum games and then apply this concept to a very interesting game called The Battle of the Bismarck Sea—a game that is even more interesting because it actually happened.

The idea of "domination" is not difficult, and if you did exercise 10.2 in the previous section, you have already seen it before. Take a look, again, at the matrix that appears in that exercise. Note the following fact: every entry in row R1 of that matrix is greater than the corresponding entry in row R2: 6 is greater than −1 and 0 is greater than −5. Now recall that each entry in the matrix represents Rose's (the row player's) outcome, and so the larger the number in the matrix, the better things are for Rose. It follows that Rose should *always* select row R1 as her strategy, since selecting R2 instead will *always* make her worse off, no matter what column Colin selects as his strategy.

Of course, Colin, who, as a rational player, will realize that Rose will play row R1, will then play column C2 as his strategy. The net result in this particular game is that no money changes hands.

This example illustrates the essence of the concept of domination. We say that one row, say R_i, dominates another row, R_j, if every entry of R_i is at least as big as the corresponding entry of R_j, and at least one entry of R_i is strictly larger than the corresponding entry of R_j. The significance of row domination is simply this: if one row dominates another, then it *never* makes sense for the row player to select the dominated row as her strategy.

There is an analogous concept of domination for columns, of course. However, remembering that the entries reflect the outcome for the row player, it follows that the better column for Colin is the one with the smallest entries. Thus, we say that one column, C_i, dominates another column, C_j, if every entry of C_i is less than or equal to the corresponding entry of C_j, and at least one entry of C_i is strictly less than the corresponding entry of C_j. If a column is dominated, it can be ignored, since the column player should never adopt that as his strategy.

For example, consider a zero-sum matrix game where each player has three strategies, and the corresponding matrix (with strategy names omitted to make things simpler) is given by

$$\begin{pmatrix} 2 & -1 & 4 \\ 3 & 1 & 5 \\ 0 & -2 & 1 \end{pmatrix}$$

It should only take a second to realize that row R1 dominates row R3. This means that it would be irrational for the row player to ever invoke strategy R3, and therefore, in analyzing this game, we can pretend it does not even exist. This leads to a new game matrix:

$$\begin{pmatrix} 2 & -1 & 4 \\ 3 & 1 & 5 \end{pmatrix}$$

in which (as is easily seen), column C2 dominates C1. So, it would be irrational for the column player to ever play strategy C1. If we pretend that C1 does not exist, we are led to the matrix

$$\begin{pmatrix} -1 & 4 \\ 1 & 5 \end{pmatrix}$$

where row R2 dominates R1 and column C1 dominates C2. Removing dominated rows and columns here yields the conclusion that the only rational strategic choice for the players in this game is for Rose to select R2 and Colin to select C2 as their strategies. The net result is that Rose wins 1. (We could have reached the same conclusion with a little less work by initially noting that in the 2 × 3 matrix above, column C2 dominated both columns C1 and C3, so both of these columns could have been removed simultaneously.)

This choice of strategies does not give either Rose or Colin the best possible outcome available to them. For example, if Rose selected R1 and

Colin selected C1, then Rose would win 2, not 1. But the point is that Colin, playing rationally, will never select C1; why would he, when by selecting C2, he can always do better? Our discussion of game theory in this chapter assumes rational behavior by the players; this may not be a good assumption for all real-world situations and has been the subject of some commentary, but we will not pursue this issue here.

The concept of domination seems so straightforward that one might question whether it actually has any relevance to real-life issues. To show that it does, we finish this section by looking at a wartime decision-making problem that actually occurred, namely the Battle of the Bismarck Sea. This battle, modeled as a game, was discussed in the 1954 Haywood paper referred to in the previous section of this book. A textbook account of this game appears in [RU], which devotes several chapters to an excellent introductory account of game theory.

The underlying facts in the Bismarck Sea case are as follows: during World War II, a Japanese commander was responsible for routing a troop and supply convoy from one location to another in the South Pacific. He had two choices: sail north of New Britain Island, or south of it. In either case, the trip would take three days. Because of weather conditions, visibility south of the island was good, but visibility north of the island was not.

General Kenney, commander of the Allied Air Forces in the Southwestern Pacific area, was ordered to intercept the convoy and inflict maximum damage on it. Good weather conditions favored him because the clearer the weather, the more effective his bombing. He also had a choice of two strategies: concentrate his reconnaissance aircraft along the northern part of the island or the southern part. If Kenney's aircraft were not located where the convoy was, there would be a delay in bombing the convoy. Based on these considerations, Kenney concluded that the number of clear days of possible bombing of the convoy could be summarized as follows:

	Japan: N	Japan: S
US: N	2	2
US: S	1	3

This is a zero-sum game where Kenney wants the most number of clear bombing days and the Japanese commander wants the least number. A glance at the numbers above reveals that this game is a dominated one: every entry in column 1 is less than or equal to the corresponding entry in column 2, and one entry (in row 2) is strictly less. So, assuming that the Japanese commander behaved rationally, he would choose to take the northern route. Knowing this, Kenney would also choose to move his planes north, so as to get two good bombing days rather than 1. In fact, this is exactly how things worked out in real life: both went north.

Exercises

10.3. Consider the 2×3 matrix game $\begin{pmatrix} 0 & 1 & 3 \\ -1 & 4 & 2 \end{pmatrix}$. Rose has two strategies, R1 and R2; Colin has three strategies: C1, C2, and C3. Which strategy should each player use? What is the outcome of the game?

10.4. It was observed in Section 10.1 that military commanders should use their prudent strategies. What outcome does that dictate for Kenney in the Battle of the Bismarck Sea?

10.3 Zero-sum Games: Saddle Points

Domination of rows and columns is useful when it exists, but it doesn't always exist. In a typical zero-sum game, it is entirely possible that no row and no column dominate each other. How might we search for good strategies in that case?

Consider, for example, a zero-sum game given by the following matrix, where we can denote the three-row strategies by R1, R2, and R3, and the three-column strategies by C1, C2, and C3. To simplify the matrix that appears, we have omitted the names of the strategies:

$$\begin{pmatrix} 2 & 3 & 4 \\ 0 & 5 & -1 \\ 1 & 0 & 6 \end{pmatrix}$$

It is easy to check (do so!) that in this matrix, no one row or column dominates another. But perhaps another technique can be used to determine a good strategy.

Recall that in Section 10.1 of this chapter we talked about a strategy that could be summed up by the phrase "hope for the best, but prepare for the worst". We also called this (for the row player) the maximin strategy or the prudent strategy. Suppose Rose were to play this strategy in this game. The worst possible results for the three-row strategies are 2, −1, and 0, respectively. The best of these worst outcomes is 2, and that occurs in row 1. This means, in other words, that Rose's prudent strategy is to play row 1. By doing so, she can guarantee that she will always win at least 2.

As for Colin, who seeks small outcomes (and for whom the worst outcomes are the larger numbers), you can check (do so!) that Colin's prudent strategy is to select column 1. In this case, it turns out that the best of the three worst possible outcomes also turns out to be 2. In other words, in this particular game, Rose's maximin = Colin's minimax (= 2). When the row player's maximin and the column player's minimax turn out to be equal, we say that the common value of these numbers is a *saddle point*. A basic principle of game theory is: *if a saddle point exists, both players should play a strategy that contains it.*

Let us try and offer some heuristic justification for this principle. First note that in the matrix above, if both players play the saddle point strategy (Rose plays R1, Colin plays C1), then neither player has any incentive to unilaterally change strategies—if, for example, Rose were to switch to R2, she would wind up with 0 instead of 2, and if Colin were to switch to C3, say, he would lose 4 instead of 2. In other words, the saddle point in this game is an *equilibrium* (a technical term that has meaning in a more general context).

No other choice of strategies has this property in this game. If, for example, Rose were thinking of selecting strategy R2 and Colin were thinking of selecting strategy C2, then Colin would benefit from switching his choice to C3. But if Rose knew or suspected that Colin would do this, she would switch her strategy choice to R3 and make 6. But if Colin suspected she would do this, he would change his mind and choose C2.... And so it goes. The point is that things don't "settle down" until the players land on 2.

As another reason why the saddle point strategy (R1 and C1) is a good one, think in terms of guarantees. We have previously noted that by using her saddle point strategy, Rose can guarantee she will win at least 2. But by invoking his saddle point strategy, Colin can guarantee that he will never lose more than this amount.

Exercises

10.5. Determine whether the matrix game $\begin{pmatrix} 2 & -1 & 3 \\ 1 & 0 & 1 \\ 3 & -2 & 0 \end{pmatrix}$ has a saddle point.

10.6. For the matrix in the previous problem, can Rose and Colin's optimal strategies be determined without using saddle points? Explain.

10.7. Determine whether the matrix game $\begin{pmatrix} 0 & 1 & 4 \\ 1 & 3 & -1 \\ 2 & 1 & 3 \end{pmatrix}$ has a saddle point.

10.4 Zero-sum Games: No Saddle Points

In the previous section, we saw that if a zero-sum matrix game has a saddle point, then the optimal strategy for the players is to play the row and column containing that point. Of course, not all zero-sum games have a saddle point: a simple example of one that does not is

$$\begin{pmatrix} 1 & 0 \\ -1 & 2 \end{pmatrix}$$

where it can be easily verified that Rose's maximin value is 0, but Colin's minimax value is 1. It can also be easily checked that there is no row or column domination. Thus, none of the methods that we have discussed to date gives us a way of determining a good strategy for either player.

In fact, this example illustrates that perhaps we need to broaden our horizons when considering just what constitutes a "strategy". Up to now, we have been thinking of a "strategy" as a plan to always select one row or one column. But can such a plan ever be optimal in this case? Suppose, for example, that Rose mentally decides to select row 1 as her strategy. If somehow Colin should become aware of that, or guess that that is what she will do, he will, of course, select column 2 as his strategy, thereby giving Rose nothing. But if Rose anticipates *that* decision, she will select row 2 as her strategy, giving her a profit of 2. But if Colin, playing mental chess and thinking many moves ahead, anticipates that decision, he will select column 1 as his strategy, giving Rose a loss (and him a profit) of 1. But if Rose plays mental chess too and is one move ahead, she will select row 1... and we are back where we started from.

So, given that the plan to always select one particular row or column is not an optimal one, how can we modify this plan? The trick is to use probability and to choose a row or column randomly in accordance with a given probability; for example, choose row 1 with a certain probability, row 2 with a certain probability, etc.

We should make it clear exactly what we mean by this. If, for example, we say, with reference to the example above, "Rose's strategy is to choose row 1 with probability ⅓ and row 2 with probability ⅔", we do not mean that Rose chooses row 1 on one occasion, then row 2 on the next two occasions, and then chooses row 1 again. The last thing in the world that Rose wants is to establish a pattern that can be identified and exploited by Colin. What we mean, instead, is that Rose uses some kind of randomizing mechanism to select a row. Perhaps she has three marbles in her pocket—one black, two white—and draws one at random before making a decision as to which row strategy to use. If she picks a black marble, she chooses row 1; otherwise, she chooses row 2. So, Rose herself does not know what row she is going to pick, right up until the time she picks it.

This, then, is our new notion of strategy: a selection of probabilities for each row (or column). The probabilities for all the rows must add up to 1, of course, and likewise for columns. If probability 1 is assigned to one particular row or column, that is called a *pure strategy*; otherwise, a strategy is called a *mixed strategy*.

We will adopt the following notation for a mixed strategy. If, say, Rose has three pure strategies corresponding to the rows R1, R2, and R3, and adopts a strategy of picking these rows with respective probabilities p, q, and r, we denote this mixed strategy (p, q, r). Similar notation will, of course, apply in the case of Colin choosing a mixed-column strategy, and in the case of more, or less, than three rows or columns.

Of course, all this leads to the inevitable question: what suggestions can we offer Rose and Colin as to what probabilities to pick for each row or column? In the next section, we will answer this question in the simplified context of 2 × 2 matrices, i.e., situations where each player has a choice of two pure strategies. But first, we will explore mixed strategies in a bit more detail.

We'll first consider as an example the matrix mentioned above:

$$\begin{pmatrix} 1 & 0 \\ -1 & 2 \end{pmatrix}$$

Since Rose is going to select each row with a given probability, she cannot possibly know in advance what her outcome will be, even if she knows what column Colin will choose. Since there are different possible outcomes, each with its own probability, the best Rose can do is determine her *expected value*, a concept that, you may recall, was first introduced way back at the end of Chapter 1.

So, for example, just to experiment, let us assume that Rose will pick a row by flipping a fair coin. This means that she selects row 1 with probability ½ and row 2 with probability ½. In our new notation, this means that Rose adopts the mixed strategy (½, ½). Now, if Colin selects column 1, this means that Rose will get 1 with probability ½ and −1 with probability ½, so her expected value is ½(1) + ½(−1) = 0. If Colin chooses column 2 then Rose's expected value is ½(0) + ½(2) = 1.

What if Colin also adopts a mixed (rather than pure) strategy and selects a column randomly with some probability? Then, it turns out, the expected value will be some number between 0 and 1. This is because a weighted average of two numbers must be between these two numbers, so a weighted average of 0 and 1 must be between 0 and 1.

For example, suppose Colin adopts strategy (⅓, ⅔). In other words, he selects column 1 with probability ⅓ and column 2 with probability ⅔. Then, since we are assuming that Rose and Colin make decisions independently, it follows that the probability that Rose plays R1 and Colin plays C1 is the product of ½ and ⅓, or ⅙. Thus, Rose wins 1 with probability ⅙. In a similar way, we see that Rose wins 0 with probability (½)(⅔) = ⅓, loses 1 with probability ⅙, and wins 2 with probability ⅓. Rose's expected value is 1(⅙) + 0(⅓) + (−1)(⅙) + 2(⅓) = ⅔.

The observation made in the preceding paragraph allows us to say, more generally, that if one player in a zero-sum game adopts a mixed strategy, then the other player's best response is a pure strategy. Here is another illustration, using the matrix

$$\begin{pmatrix} 1 & -1 & 2 \\ 2 & -2 & 1 \\ 0 & 1 & 3 \end{pmatrix}$$

An Introduction to Game Theory

As you can check, Rose's maximin value is 0. This means that she can arrange, by selecting row 3, that she at least comes out even in this game. However, Colin, on the other hand, has minimax value 1, which means that he can keep from ever having to pay Rose more than 1. If both players play their prudent strategies, Rose will play row 3 and Colin will play column 2; Rose will win 1. Since the minimax and maximin values are not equal, however, Rose and Colin may wish to consider using mixed strategies.

Suppose Rose decides to play each row with equal probability—i.e., probability $1/3$. If Colin were to somehow become aware of Rose's plan, what would his best response be? We know from the previous comment that his best response (from the standpoint of expected value) is a pure strategy. If (check these claims!) he chooses column 1, Rose's expected value would be 1; if he chooses column 2, Rose's expected value would be $-2/3$; if he chooses column 3, Rose's expected value would be 2. Clearly, therefore, if Colin became aware that Rose was going to use the mixed strategy $(1/3, 1/3, 1/3)$, he should play column 2. This is Colin's best response to Rose's mixed strategy, but Rose's mixed strategy is certainly not her best response to Colin's pure "column 2" strategy. One thing we might ask is whether Rose and Colin both have mixed strategies, each of which is the best response to the other. This would be a mixed-strategy analog of the kind of "equilibrium" situation where a saddle point exists. We look at this issue in the next section, but for the sake of simplicity, we confine ourselves in that section to the case of 2 × 2 matrix games.

Exercises

10.8. For the matrix game $\begin{pmatrix} 0 & 1 & 4 \\ 1 & 3 & -1 \\ 2 & 1 & 3 \end{pmatrix}$ that Colin becomes aware that Rose is going to use the mixed strategy $(1/3, 1/3, 1/3)$. What is his optimal counter-strategy?

10.9. In the previous question, if Colin uses mixed strategy $(2/3, 0, 1/3)$, what is Rose's best counter-strategy?

10.5 Solving 2 × 2 Zero-sum Games

We have previously seen that if a zero-sum game has a saddle point, then selecting a row and column containing that saddle point gives a pair of strategies for Rose and Colin that are in equilibrium, in the sense that neither party has an incentive to unilaterally change strategies. Put another way, each strategy is the optimal response to the other. In this section, we will identify comparable equilibrium mixed strategies for both players in the case where no saddle point exists. However, we will work in the case of 2 × 2 matrix games, where the determination of these strategies is particularly simple.

The best way to identify these strategies is to work with an example. So, consider a 2 × 2 game with two-row strategies R1 and R2 and two column strategies C1 and C2, given by the matrix

$$\begin{pmatrix} 2 & 0 \\ 1 & 3 \end{pmatrix}$$

Searching for optimal strategies, we first look for row and/or column domination but quickly see there is none. We next look for a saddle point, but come up empty here as well: Rose's maximin value is 1; Colin's minimax value is 2. (This information does allow us to at least conclude that Rose can always guarantee that she will win at least 1, and Colin can guarantee that she never wins more than 2.)

So, we look for a good mixed strategy for Rose. Let us denote it, using our previous terminology, by $(p, 1-p)$. Here, p denotes the probability of selecting R1; determination of that number will specify our good strategy for Rose. We know that Colin's best response, as far as Rose's expected value is concerned, is a pure strategy. If Colin selects C1 as his response, it is easy to see that Rose's expected value is $2p + 1(1 - p) = p + 1$; let us denote this as E1. Likewise, if Colin selects C2 as his response, it is easy to see that Rose's expected value is $0p + 3(1 - p) = 3 - 3p$. Let us denote this expression by E2. Now, what can we say about these expected values? Which is better for Rose?

Obviously, this answer depends on p. If, for example, $p = 0$, then E1 < E2, but when $p = 1$, E1 > E2. So, at some point, E1 switches from being less than E2 to being greater than E2. This occurs when E1 = E2, and a little algebra shows that $p + 1 = 3 - 3p$ when $p = ½$. So, our "equalizing" approach leads to the mixed strategy (½, ½) for Rose. For this strategy, her expected value will be 1½ whether Colin chooses column 1, column 2, or any mixed strategy of column 1 or column 2. She can thus guarantee herself an expected long-run profit of at least 1½, no matter what strategy Colin plays.

On the other hand, if she plays a different strategy $(p, 1 - p)$, then it is possible that, depending on what strategy Colin plays, her expected value might be less than 1½. For if $p < ½$, Colin's best possible counter-strategy is column 1 (because for such p, E1 < E2) and $1 + p < 1½$, if $p > ½$, Colin's best possible counter-strategy is column 2 (because for such p, E1 > E2) and $3 - 3p < 1½$.

We emphasize that we are not saying here that selecting this strategy will always guarantee Rose her best possible outcome. But from a defensive point of view, this is Rose's optimal strategy: it gives her the best possible *guaranteed* expected value.

Of course, determination of a good strategy for Colin follows precisely the same reasoning. Let us work through the details. If we denote Colin's mixed strategy by $(q, 1 - q)$, with q to be determined, we see that his expected payouts to Rose are $2q$ if Rose plays row 1 and $q + 3(1 - q) = 3 - 2q$ if Rose plays row 2. Equating these gives $q = ¾$. This produces an expected payout to Rose of 1½ no matter what row Rose selects as her pure strategy. And, just as before, it is not hard to see that this is the best possible guaranteed expected payout Colin can hope for. We leave this for the reader.

Note also that the two expected values are the same: 1½. So, Colin and Rose's "equalizing" strategies turn out to yield the same guarantees for the players. Again, this is not a coincidence. This common value is called the value of the game; if positive, this means that the game favors Rose, who can expect a guaranteed profit in the long run. Likewise, if negative, the game favors Colin.

The two strategies that we have arrived at here—one for Rose, one for Colin—are an example of what is called a *Nash equilibrium* for the game, a concept named after John Nash, the subject of the book and movie *A Beautiful Mind*. This concept is of considerable importance in game theory, but we will content ourselves with this brief description of the idea.

The idea of selecting a mixed strategy that "equalizes" the expected values of all possible pure strategies of your opponent is not limited to 2×2 games, but in larger games, the algebra becomes more complicated. One game where it is manageable, however, is the game of Rock/Paper/Scissors, discussed in exercise 10.1. If rows 1, 2, and 3 denote, respectively, "rock", "paper", and "scissors", and similarly for columns 1, 2, and 3, then the matrix for the game is

$$\begin{pmatrix} 0 & -1 & 1 \\ 1 & 0 & -1 \\ -1 & 1 & 0 \end{pmatrix}.$$

If we denote an arbitrary mixed strategy for Rose by (p, q, r), then her expected values (if Colin plays columns 1, 2, and 3, respectively) are

$$q - r$$
$$r - p$$
$$p - q$$

and if we set these three expressions equal to each other, we get

$$q + p = 2r$$
$$r + p = 2q$$
$$r + q = 2p$$

which, upon subtracting the second equation from the first, yields $q = r$. Replacing q by r in the first equation gives $p = r$. So, $p = q = r$, and, since the sum of these numbers must be 1, their common value must be ⅓. In other words, Rose should pick either rock, paper, or scissors randomly, with equal probability. Interestingly enough, this is the same strategy that James Bond thought of in *You Only Live Twice*.

Exercises

10.10. Consider the zero-sum game given by the matrix $\begin{pmatrix} 2 & 0 \\ 1 & 3 \end{pmatrix}$. If Rose chooses the "equalizing" strategy discussed in this section, what strategy will she adopt? What will her expected long-term winnings be?

10.11. Redo the previous problem, this time finding Colin's "equalizing" strategy and his expected payout to Rose.

10.12. Consider the zero-sum game given by the matrix $\begin{pmatrix} 0 & 1 \\ 2 & 1 \end{pmatrix}$. If Rose chooses the "equalizing" strategy discussed in this section, what strategy will she adopt? Is this the best thing for her to do? Explain.

10.13. Consider the zero-sum game given by the matrix $\begin{pmatrix} 1 & 3 \\ 0 & 2 \end{pmatrix}$. What happens if Rose attempts to determine the "equalizing" strategy discussed in this section? What accounts for this?

10.14. Find the "equalizing" strategies for Rose and Colin for the zero-sum game given by the matrix $\begin{pmatrix} 0 & 2 \\ 1 & -1 \end{pmatrix}$. Does the game favor Rose or Colin?

10.15. Same as the previous exercise, this time for the game given by the matrix $\begin{pmatrix} 3 & -2 \\ -1 & 1 \end{pmatrix}$.

10.6 A Simplified Poker Game

In their book [HK], Herstein and Kaplansky discuss a 2 × 2 zero-sum matrix game that models the concept of bluffing in poker. It is a little different than the games we have been discussing to date because Colin's strategic choices depend on what Rose does, but what we have covered up to this point allows us to discuss it.

Here are the rules. Rose and Colin are each randomly dealt a card bearing the number 1 or 2. They are equally likely to get either number. Neither player knows the other's number. Rose makes the first move, and she does so by saying either "High" or "Low". If she says "Low", the players compare cards and the person with the high card wins $8 from the person with the low card. If she says "High", then Colin must either Call or Fold. If he folds, he pays Rose $8. If he calls, the players compare cards, and the one with the higher number wins $12 from the other. If, in any case, the numbers are the same, no money changes hands.

An Introduction to Game Theory

In theory, Rose has four strategies:

Call High on 1, High on 2
Call High on 1, Low on 2
Call Low on 1, High on 2
Call Low on 1, Low on 2

On the other hand, it is obvious that Rose should never call Low on 2, since she can always do better by calling High in this case. (In the language of this chapter, this strategy is dominated.) So, we can eliminate the second and fourth strategies above, leaving only two: "Call High on 1, High on 2" (R1) and "Call Low on 1, High on 2" (R2). Of course, strategy R1 amounts to bluffing on the part of Rose.

As for Colin, he too has four possible strategies:

Call on 1, Fold on 2
Call on 1, Call on 2
Fold on 1, Fold on 2
Fold on 1, Call on 2

Here again, though, we can eliminate two of these strategies, specifically the first and third, because it never makes sense for Colin to fold when holding a 2; clearly, if he calls in this case, he cannot lose. So, we can assume Colin has two strategies: "Call on both 1 and 2" (C1), "Fold on 1, Call on 2" (C2).

Now we need to fill in the entries of the matrix. We cannot say with certainty what Rose will get if, say, she plays R1 and Colin plays C1, because this depends on what cards she and Colin are holding. But we can determine her expectation. First, we consider the four possible scenarios and determine the outcome (to Rose) in each one:

Rose holds 1, Colin holds 1. Outcome: 0
Rose holds 1, Colin holds 2. Outcome: −12
Rose holds 2, Colin holds 1. Outcome: 12
Rose holds 2, Colin holds 2. Outcome: 0

Each of these scenarios occurs with probability ¼, so Rose's expectation is ¼(0 − 12 + 12 + 0) = 0. This becomes the upper-left entry in our game matrix.

Now let us assume Rose plays strategy R1 and Colin plays C2 (he folds if he holds a 1). The four possible scenarios and their outcomes for Rose now become:

Rose holds 1, Colin holds 1. Outcome: 8
Rose holds 1, Colin holds 2. Outcome: −12
Rose holds 2, Colin holds 1. Outcome: 8
Rose holds 2, Colin holds 2. Outcome: 0

Rose's expectation is, again, one-fourth the sum of these outcomes, which now is 1. So, at this point, we have half the matrix of this game filled in:

$$\begin{pmatrix} 0 & 1 \\ * & * \end{pmatrix}$$

The entries marked * must yet be determined. We leave it to the reader to (using similar arguments) do this, getting

$$\begin{pmatrix} 0 & 1 \\ 1 & 0 \end{pmatrix}.$$

Clearly, there is no row or column domination in this matrix, and there is no saddle point either. But we can come up with a good strategy for Rose and Colin by using the method of the preceding section. It is very easy to see (check this!) that the "equalizing strategy" is (½, ½) for each player: i.e., Rose and Colin should randomly bluff half the time.

A few comments about this game: first, it clearly favors Rose (her expected value is ½). Can you give an intuitive explanation for this? Second, the game matrix obviously depends on the amounts at stake. If, for example, we increase the penalty for bad bluffing (i.e., change $12 to a larger amount), then bluffing becomes less desirable, and the optimal strategy for each player would be to bluff less than half the time. We explore this idea in the exercises.

Exercises

10.16. Can you, as asked in the text, give an intuitive explanation for why this game favors Rose?

10.17. Consider the game above, but this time assume that if Rose says "High" and Colin calls, the person with the larger card wins $20 instead of $12. Using the same reasoning as above, determine the new game matrix.

10.18. (Continuation of previous question.) Find the "equalizing strategies" for Rose and Colin under the new game matrix. Explain the difference in outcome.

Index

A

Addition principle, 47–48
Average outcome, 12

B

The Battle of the Bismarck Sea, 90, 94
Big 8 bet, 44–45
Big Red bet, 45
Binomial coefficients, 54–55, 74
Birthday problem, 51–52
Blackjack game
    calculations, 77–79
    rules of, 75–76
Bonus points, 82
Bringing Down the House (Mezrich), 80

C

Candy Land game, 89
Card counting, 80–81
Card games, 3
Carson, Johnny, 52
Certain event, 6
Code words, 50–51
Complement of an event, 10–11
Conditional probability, 26–29, 31
Counter-intuitive solution, 8, 27, 36, 51
Counting rules, 47–52
Craps, 28
    casino, 44–46
    rules, 39–41
Cuban Missile Crisis, 90

D

Domination, 92–94
Don't pass bet, 44

E

Elementary events, 5–6
Elements of set, 1

Equalizing strategy, 104
Even- or odd-money bets, 18, 21, 23, 46
Events
    certain, 6
    definition, 6
    elementary, 5–6
    impossible, 6
    odds of, 16
    probability of, 7–9
Expected value, 12–14, 20–22
    in gambling, 13
    negative, 13–14

F

Farkle
    probability of farkling, 84–88
    rules of, 82–83
Flipping of coin, 4–5, 9
Free Odds, 45–46

G

Game theory, 62, 89
    strategic choice, 91
Geometric progression, 23

H

High-Low card counting method, 80–81
House odds, 16

I

Impossible event, 6
Independent events, 33–35
Insurance bet, 76, 78
Intersection of sets, 2
Iowa Lottery, 69

K

k-element subsets of an n-element set, 54
Kenney, General, 94
Keno, 72–74

L

Law of Large Numbers, 9
Law of Total Probability, 30–32
Let's Make a Deal, 36
Lotteries, 68

M

Martingale system, 23–24
Mathematical odds, 16
Matrix, 90
Maverick Solitaire, 61
Maximin strategy, 91
Military Decision and Game Theory (Haywood), 91
Mixed strategy, 97
Monty Hall Problem, 36–38
Multiplication principle, 4, 48–49

N

N factorial, 49
Non-independent events, 35
Null set, 2, 6

O

Odds of an event, 16
Ordered list, 54

P

Pass bet, 44
Pat hand, 61
Payout rate, 45
People v Collins, 35
Permutations and combinations, 53–56
Pick 10 bet, 73
Pick 6 game, 72
Poker game, 103–104
　flush, 59
　four of a kind, 59
　full house, 59
　nothing, 61
　one-pair hand, 61
　poker hands, 58
　royal flush, 58–59
　straight, 60
　straight flush, 59
　three of a kind, 60
　two-pair hand, 60
Positive integers, 1
Powerball game, 69–71
Probabilities: The Little Numbers that Rule Our Lives (Olofsson), 78
Probability
　conditional, 26–29, 31
　of farkling, 84–88
　Law of Total Probability, 30–32
　of no-repeat birthdays, 51–52
　of winning, 41–43, 69–70
Probability of events, 7–9, 26
　counter-intuitive results, 8–9, 27, 36, 51
Prosecutor's Fallacy, 29
Prudent strategy, 91
Pure strategy, 97

R

Rank of a card, 3
Rolling of die, 4, 6, 82–83
　expected value of, 13
Roulette game, 18–20
　calculations, 20–22
　even- or odd-money bets, 18
　roulette table, 19
　roulette wheels, 19–20
Roulette "systems," 23–24
Rules of the game
　Blackjack, 75–76
　craps, 39–41
　farkle, 82–83

S

Saddle points, 95–96
Sample spaces, 3–4, 6, 28–29
　size of, 4
Savant, Marilyn vos, 36–37

Index

Set theory, 1, 50
Shooter's game, 39–43
Strategic decisions, 89
Subset, 2
Sudden Infant Death Syndrome (SIDS), 34–35
Suit of a card, 3

T

Texas Hold 'Em, 66–67
Thorp, Edward, 80
True odds, 16
2 × 2 game, 99–104
Two-person zero-sum games, 91

U

Union of sets, 2
Unordered list, 54
The Untouchables (TV show), 69

V

"Very close to" notion, 9
Video poker, 62–65
    table of payouts, 63

W

War game, 89
Weighted average, 20

Y

You Only Live Twice, 103

Z

Zero-sum game, 90
    concept of domination, 92–94
    2 × 2 game, 99–104
    no saddle points, 96–99
    saddle points, 95–96
    zero-sum matrix game, 93–94